PHILOSOPHY, EVOLUTION AND HUMAN NATURE

PHILOSOPHY, EVOLUTION AND HUMAN NATURE

Florian von Schilcher
and
Neil Tennant

ROUTLEDGE & KEGAN PAUL
London, Boston, Melbourne and Henley

First published in 1984
by Routledge & Kegan Paul plc
39 Store Street, London WC1E 7DD, England
9 Park Street, Boston, Mass. 02108, USA
464 St Kilda Road, Melbourne
Victoria 3004, Australia and
Broadway House, Newtown Road,
Henley-on-Thames, Oxon RG9 1EN, England
Set in IBM Press Roman by Columns, Reading
and printed in Great Britain by
Redwood Burn Limited, Trowbridge, Wilts.

Library of Congress Cataloguing in Publication Data

Schilcher, Florian von, 1944-

Philosophy, Evolution, and Human Nature.
Bibliography: p.
Includes index.
1. Evolution. 2. Human Evolution. 3. Biology--
Philosphy. I. Tennant, Neil. II. Title.
OH371.S38 1983 575 83-16112

ISBN 0-7100-9767-0

CONTENTS

PREFACE

The idea of writing a book of this kind first occurred to us in the spring of 1977. NWT was visiting FvS deep in the Bavarian woods and we were soon caught up discussing the mental life of the latter's dogs. As with any decent topic in philosophy, the ramifications brought us to an impasse: one typical of confrontation between a hard-nosed scientific reductionist and an analytical philosopher. To resolve this disagreement and others like it we agreed to work together on this book. It represents a shared view after subsequent months of lively discussion.

The major part of our research together was done during 1979 when NWT was able to visit Germany on Alexander von Humboldt and British Academy Fellowships. To both these organisations, he expresses his gratitude. He is also grateful to Professor W. Stegmüller for the benefit of formal association with the Institut für Wissenschaftstheorie und Statistik in Munich.

By the end of 1979, the plan of the book was beginning to take shape. We wished to address ourselves to both philosophical problems about biology and biologically informed views on standard philosophical problems. Several passages ancestral to present sections in the first two chapters were written, feverishly, in idyllic conditions in Pontresina in the valley where Nietzsche went mad. Later in a coastal farmhouse in County Waterford in Ireland, the seeds of emergent organisation in the material were sown. In the summer of 1981, a final month's work in Dietramszell saw a working draft complete. But for this period of sustained hospitality from Isabella von Schilcher, the book might not have been written.

There followed over a year of critical comments, two-way translations in German and English, stylistic corrections and further arguments over details. Among friends and scholars who have been good enough to discuss aspects of the subject with us and read parts of the work and offer their criticisms, we would like to mention: M. Dow, J. Hall, J. Jacobs, J. Kiefer, A. Manning, J. Maynard Smith, W. McGrew, K. Moritz, M. Pastrana, A. Peacocke, and G. Roth. Among the exchanges that have inspired us, especially in the writing of chapter IV, we cannot resist recording the following:

NWT: Kate, how do you know that you think with your head?
Kate (aged 6): Because when you think, it comes out of your mouth in words, and you know that water can't run uphill.

We have also benefited greatly from discussions of extracts presented to the following audiences:

Newman Circle, Edinburgh
The Philosophy Societies at the Universities of Edinburgh, Glasgow, Stirling, University College (London), St. Andrews
Bradford Inter-Disciplinary Seminar in Human Studies
The University of London Philosophical Forum
Department of History and Philosophy of Science, Cambridge
Epistemics Research Seminar in Edinburgh
Thyssen Group in Philosophy
The International Wittgenstein Colloquium in Kirchberg, 1982
Cambridge University Press kindly granted permission to use material from Tennant (1983).

Thanks to Linda Greig, the book was able to evolve through successive drafts; only through her cheerful efficiency did it find its way into print at all.

We dedicate this book to Megan, Elizabeth and Kate, and to Claudia, Fabian and Anna.

Neil W. Tennant, Stirling
Florian von Schilcher, Dietramszell

INTRODUCTION

This book is aimed at an audience of both philosophical and scientific temper. For professional scientists and philosophers it may half the time be telling twice told tales. But because of this dual purpose, we hope and intend that the educated layman concerned with human nature in a modern world of science will derive some pleasure and benefit in reading it. We would like to think that, by contributing our own expertise in this way, we have brought our disciplines into more fruitful contact.

Our chosen subject is so wide that it would be impossible for any inter-disciplinary team of specialists to cover it in a book length study of completely satisfactory breadth, depth and cohesion. The minimum we wish to have achieved is to have explained clearly the details of modern evolutionary theory and assessed its consequences for a theory of human nature whose main concerns are morality, knowledge, mind and language. One of us by training is a logician and philosopher; the other a behaviour geneticist. Together we have agreed on the philosophical framework for the ideas developed in this book.

The overall project to which it is but a small contribution is to give a unified philosophy and science of nature, and account for man's place within it. The project is materialist in spirit. It may even be nominalist, in eschewing reference to abstract entities. But it allows so-called anomalism of the mental and social. That is to say, it accepts the claim that there is nothing but the material, or physical in nature; but that while all physical facts suffice to determine all that is the case, nevertheless the language of the mental and the social may be ineliminable in certain kinds of explanation, certainly for heuristic and possibly even for logical reasons. We shall not, however, be pursuing the question of nominalism; for whether scientific theories can eschew all reference to abstract objects is at once a more specialised and more fundamental problem for the philosophy of science as a whole.

Scientific theories are not all strongly predictive, and do not all deliver deterministic laws. In the case of evolutionary theory in particular, one is more concerned with inference to the best explanation. In this context man, as a physical product of genes and a highly developed intentional system, is the most problematic natural kind of all.

His greatest gift, language, is bound up both empirically and conceptually with intelligence, creativity, rationality, morality, consciousness and personal and cultural identity. Overly strong emphasis on the conceptual links with language, on how our linguistic capacities are constitutive of or preconditions for these various others, tends to deepen the divide between man and the rest of the evolved natural order. We do not go into any of the paleontological details, but start with the scientific consensus that Darwin's case for our descent is proven. Our task is then to assemble the conceptual rungs of the evolutionary ladder that we have climbed and that uniqueness fetishists would kick away, and put them together in a plausible order.

In so doing we have concentrated on three main areas: sociobiology, evolutionary epistemology, and our own evolutionary scenario for language that draws on a variety of disciplines ranging from neurobiology to theoretical linguistics. Given the potential cast of this net, a comprehensive review of the literature is impossible. But we have tried to let only little fish slip through. We sometimes refer only to certain authors on topics ably discussed by many others. Moreover, we have had to absorb so much in the way of background reading that by the time of writing the ancestry and alchemy of ideas was often lost or buried.

This book seeks to emphasise the pervasive importance of biological considerations for one's understanding of human nature. We are aware that biological thinking has had noxious political associations in the past. We wish emphatically to dissociate ourselves from any attempt to press a conception of human nature such as ours into the disservice of any individual or group. But we believe the serious minded and well intentioned should come to grips with the problem of human nature in the available light of modern science and analytical philosophy, and challenge abiding misconceptions about human nature, howsoever congenial these may be to commendable charters of human rights and freedoms. Perhaps the value of the fact-value distinction lies precisely in the facts.

In writing a book on the philosophy of biology, one finds that the two disciplines bear on each other in two different ways. First, biology raises certain problems within the philosophy of science. These concern evidence and method, concept and theory formation, prediction and explanation, and reductionism. With these we are mainly concerned in the first chapter, whose aim is to present a clear account of evolutionary theory as it is later to bear on our philosophical preoccupations.

Secondly, biology throws light on traditional philosophical problems in an interesting way. The second chapter explores the connection between sociobiology and ethics, distinguishing evolutionary ethics from evolutionary theories about ethics. The third chapter examines evolutionary epistemology, the boldest attempt to date to naturalise the theory of knowledge, to make it a part of science rather than a precondition for it. We discuss the exegetical credentials claimed by evolutionary epistemologists in Kant, and the account of mind, objectivity, realism, perception and cause – in brief, the metaphysical underpinnings – required by this programme. Finally the fourth chapter addresses itself to the nature of human language and the behavioural and neurological foundations of its likely evolution. An appropriate metaphor that emerges is that of an evolutionary spiral within which stepwise incremental gains in syntax and semantics derive from increasingly complex behaviour – at first contextual, later conventional. Meaning is parasitic, from the evolutionary point of view, upon movement and noise.

No picture of man is complete without a Darwinian frame. We hope to have blocked in the major areas of the canvas. No competing theory of human nature – such as existentialism, Marxism, creationism, narrow behaviourism or Freudianism, to name but a few – addresses itself in the same way to the full range of objectively observable truths about man with the same integrative thrust in search of explanation.

I

THE MODERN THEORY OF BIOLOGICAL EVOLUTION

1 EVOLUTION BY NATURAL SELECTION

A species of moth somewhere in industrial Britain was once characteristically light coloured. Among them, dark mutants occasionally recurred. Then smog from new factories nearby darkened the trees they lived on. Conspicuous against the darker background, the light ones were easy prey for birds. After a few generations almost all the moths were dark, and no longer such easy pickings. Single individuals though, did not darken with age. Rather, each generation had proportionally more dark members than the preceding one.

How did the moths thus adapt to the new conditions? Most importantly, because within each generation they varied in their colouring. The dark ones, being better camouflaged, left more offspring than the light ones. These offspring inherited the dark colouring of their parents, and in turn left more offspring than the other, light moths of their generation. And so the famous phenomenon of industrial melanism was observed (Kettlewell, 1961). Evolution by natural selection led to an adaptive change in colouring. This example strikingly illustrates what lies behind evolutionary change. To put it quite generally, evolution takes place if the following three conditions are fulfilled (Lewontin, 1969):

1 There is *phenotypic variation*. Individuals differ with respect to those properties whose distribution within the population the process of evolution is to change.

2 This phenotypic variation has a *genetic basis*. Offspring take after their parents regardless of the circumstances in which they grow up.

3 There is *differential reproduction*. Some parents produce more offspring than others.

If these major conditions are met then evolution follows as a logical consequence. The distribution of *phenotypes* in a group of creatures will change. By phenotype we mean, in agreement with common usage, a cluster of perceptible properties, such as height, weight, hair colour, number of digits and so on. We include in the phenotype all properties apart from the information content of the genetic material. This information content is called a *genotype*. Each individual has a genotype which, with the circumstances in which the individual grows, determines the phenotype. Underlying the changing distribution of phenotypes is, of course, a changing distribution of genotypes.

The change in gene frequencies that underlies the evolutionary process can occur quite by chance, and without 'direction'. In this case, however, the change in gene frequencies does not give rise to adaptive evolution and is not a result of natural selection. By contrast, adaptive change results from gene frequency changes in response to those conditions that are important in determining which individuals produce relatively more offspring. In Darwin's words (1859):

> any being, if it varies however slightly in any manner profitable to itself, under the complex and sometimes varying conditions of life, will have a better chance of surviving, and thus be *naturally selected*. From the strong principle of inheritance (our second major condition), any selected variety will tend to propagate its new and modified form.

In this way says Darwin, 'the innumerable species inhabiting this world have been modified, so as to acquire that perfection of structure and coadaptation which justly excites our admiration'. Darwin, of course, knew nothing of genes and gene frequencies, but clearly grasped the overall logic of evolutionary change. According to the synthetic theory, adaptation resides in a correlation between genotypes and the conditions of life. The reproductive success of dark moths was *on account of* their dark colouring. Thus darkening across the generations was an adaptive evolutionary change. It resulted from natural selection, the interplay between chance and necessity, not from mere chance events alone. Directed gene frequency changes can result in the gradual change of a species or its splitting up (micro-evolution). Natural selection can also cause species to diversify so widely as to create new genera, families or even phyla (macro-evolution). Thus natural selection gives rise not only to the often striking form, function and design of living creatures, but also to their enormous diversity.

The first step in the evolution of organic diversity leading to different higher groups of organisms such as genera, families, orders or phyla is always the emergence of a new species. Population geneticists have defined a species as a group of organisms that share a common genepool; that is, organisms that crossbreed. As soon as two groups no longer share a common genepool one speaks of two different species. This definition is agreeably objective. It does, however, have certain disadvantages. It does not apply to organisms which do not reproduce sexually. For such organisms – and they are not only lower ones – the concept of a common genepool makes no sense whatsoever. Furthermore, the definition does not apply to organisms that are not contemporaries or neighbours. In these cases one resorts instead to criteria of morphological diversity. These are much more arbitrary and controversial. Moreover, large morphological differences can accrue with time in one line of organisms which has never split up. Then one might be tempted to speak of different species, when by contrast all the change has taken place within one and the same species, in the population geneticist's sense.

We do not wish to pursue this discussion further here because it does not yield any insights into evolutionary theory that would interest the non-specialist. The imprecision of the concept has no epistemological repercussions for Darwinism. Asexual 'species' are in the minority and do not evolve in the same way as sexual ones. For them all categories of classification above the individual are imprecise, just as they are for sexual species above the species level. These categories are important only for those who seek to reconstruct the course of events that led from one or few species several billion years ago to the many millions of species that have since inhabited the earth. We are more concerned with the logical structure of Darwin's theory.

Our three major conditions guarantee evolution according to Darwinian principles provided also that the following more or less trivial conditions obtain:

1 Every new born individual has considerably fewer parents than there are members of the parents' generation. It is obvious that there could be no evolution in the sense of gene frequency changes if all individuals had to participate genetically in the creation of new offspring. Equally obvious is the fact that this condition is met by all known species.

2 Although it is logically possible to have evolution in the sense we have defined in an ever-growing population of immortal individuals, in earthbound evolution individual mortality and competition for finite resources are very important. Thus Darwin (1859, p. 68) 'many more individuals of each species are born than can possibly survive; and . . . consequently, there is a frequently recurring struggle for existence'.

3 The mechanism of heredity should be a particulate, rather than a blending one. As had been observed before Mendel's genetic discoveries were properly assimilated, only particulate heredity could maintain genetic variability (once it had arisen) long enough to make long-term evolution possible. With blending heredity, offspring values of heritable traits always lie betweeen those of the parents. Thus further mating among those offspring progressively reduces representation of the extremes. As Carlson's historical study (1966) of the gene concept shows, it took some time before blending inheritance was ruled out. Darwin's studies of complex characters in pigeons suggested some kind of blending. The first generation offspring of crosses are intermediate between parental extremes. The second generation increases only slightly in variability. It does not split into the two parental types in the way to be expected, were one gene responsible for the trait in a particulate fashion. It was also clear, however, that observed mutations would not maintain enough variability for effective selection if the hereditary mechanism were blending. This dilemma led Darwin at one time to accept a Lamarckian mechanism allowing changes in the hereditary material of blood cells to feed back into the germ cells, thereby increasing their variability. (See also Dobzhansky (1970) for a history of these views.)

4 But even with particulate heredity there must be ways of creating ever new variations since many are lost in evolution, through natural

selection and historical accidents. The main sources of new variation and mutation and recombination.

5 Besides these biological preconditions there are some physical ones that should be observed as scarcely to deserve mention: a temperate range that allows sufficiently stable macromolecules to exist, an atmosphere that lets through energy from the sun without a greenhouse effect and so on.

The regularities of natural selection that we have just sketched appear to be quite simple. Yet they form the theoretical basis of the biological sciences.

It is sometimes held that the founder of evolutionary thought, of the idea that species have changed and differentiated over long periods of time, is Jean Baptiste de Lamarck. In fact, the idea had been entertained by the ancient Greeks. Charles Darwin, however, is particularly celebrated for having proposed natural selection as the mechanism of evolution. Darwin also had his predecssors; his own grandfather Erasmus Darwin had expressed similar ideas more or less explicitly. The original publication in 1858 under the title *Evolution by Natural Selection*, however, bore alongside Darwin's name that of Alfred R. Wallace. Wallace had made the same proposal independently and had sent a manuscript containing his ideas to Darwin for criticism. At the time Darwin's thoughts had been yellowing for several years in his desk drawer; and he felt impelled to go into print with Wallace, on the advice of his mentor Sir Charles Lyell, the great geologist.

Darwin had delayed publication. He wanted to produce a conclusive argument, one backed by a vast reservoir of facts and observations. Precisely because of his highly successful collection of data supporting his theory, the publication in 1859 of the *Origin of Species* ensured that his name almost exclusively would be associated with this intellectual revolution – a revolution in our understanding of the world and of ourselves. The thought took hold that we are in all essentials more closely related to apes than to angels. It became clear that organic purpose was the product of efficient causes, and not shaped by final causes operating from the future. Aristotle's 'causa finalis' was to be wholly eclipsed by a 'causa efficiens' and godly powers by worldly ones.

Darwin's original formulation of his theory is naturally not as precise as the modern one we have begun to explain, because he had practically no knowledge of genetics. Moreover, he was not altogether clear about the distinction between conditions that are necessary, sufficient, or merely favourable for evolution. For example, he ascribed enormous importance to overproduction, the capacity of most organisms for geometric increase in numbers. This, however, helps evolution along only in so far as it creates much stronger selective pressures, thereby allowing only a small proportion of individuals in each generation to survive. The overproduction of offspring is, however, neither necessary nor sufficient for the evolutionary process. Darwin's emphasis on overproduction was prompted by more than his observations in the wild during the famous voyage of the *Beagle* to species-rich South America. He had also been crucially influenced by Malthus's lectures on the growth of populations, in which the geometric increase of populations in a world of limited resources had been heavily emphasised. This was decisive for Darwin's discovery of the law of natural selection (for whose history see Mayr (1972) and Ruse (1979a)).

There can be no doubt as to the logic behind the law. It has the form that *if* three major conditions are met, *then* evolution (in a suitably defined sense) takes place. In the world as we know it the antecedent conditions do indeed hold – they are simple synthetic truths. Individuals differ, offspring take after their parents, and some parents produce more offspring than others. In accordance with the law, evolution therefore takes place. The possible corollaries of this, namely adaptation, speciation, and so on, are manifest to the observer. But our world is merely one among all possible worlds, in all of which the law of natural selection, as a *conditional* statement, is true. The whole *conditional*, in other words, is a logical, or analytic, truth.

At the very most the sceptic could take issue with Darwin over his contention that these selection conditions are widely satisfied in the plant and animal kingdoms. He could also try to deny that what we regard as remarkable adaptations among different creatures to their environments are indeed what they are claimed to be, namely the result of natural selection alone (Reed, 1981; Gould and Lewontin, 1979). Furthermore, the sceptic could deny that the plethora of species on earth today originated through natural selection from a smaller number of ancestral species or indeed even from one.

The truth of all these claims contested by the sceptic is a legitimate object of empirical enquiry: of conjecture and explanation. It is not our purpose in this book to essay the monumental and painstaking work of taxonomists, comparative embryologists, paleontologists and molecular biologists, to mention but a few of the areas of biology that have all contributed a solid mass of evidence in support of Darwin's theory (see, for example, Mayr, 1963; Barnett, 1958; Dobzhansky *et al.*, 1977). These findings, in our view, effectively disarm the sceptic who denies that either the antecedent or the consequent of the law of natural selection holds in our world. Later we shall answer sceptical criticism of another kind, such as that of the early Popper, as to whether the law of natural selection is scientifically testable, or should be dismissed as a useless tautology.

Darwin's revolution was philosophically important because of the discredit it brought upon the platonic theory of forms. Plato's eternal and unchanging forms, in which particular creatures more or less imperfectly participated, had dominated neoscholastic thought for centuries before. The very title of Darwin's book, implying that species had had a beginning and were not immutable, was to shake one of the most fundamental assumptions of this tradition.

> That the combination of the very words origin and species embodied an intellectual revolt and introduced a new intellectual temper is easily overlooked by the expert. The conceptions that had reigned in the philosophy of nature and knowledge for two thousand years, the conceptions that had become the familiar furniture of the mind, rested on the assumption of the superiority of the fixed and final; they rested upon treating change and origin as signs of defect and unreality. In laying hands upon the sacred ark of absolute permanency, in treating the forms that had been regarded as types of fixity and perfection as originating and passing away, the *Origin of Species* introduced a mode of thinking that in the end was bound to transform the logic of knowledge, and hence the treatment of morals, politics, and religion.

This magisterial prose is Dewey's (1910). Another American scholar, Andrew Dixon White (1896), had described the *Origin* as coming 'into the theological world like a plough into an anthill. Everywhere those thus rudely awakened from their old comfort and repose had swarmed forth angry and confused.'

As it happens, theologians looking for gaps to drive in the wedge of mysticism have not been short of agreeable cracks in the Darwinian edifice. Most importantly, Darwin never said anything about the origin of life itself and his theory makes no claims about it. It is an imponderable initial condition for an otherwise well orchestrated story.

Here one might believe that godly intervention took place – although to be sure there are modern theories that attempt to push Darwinism right into the prebiological era in accounting for the origin of life. Another chink that anti-Darwinians have been keen to descry is the alleged 'missing link' between the apes and man. Teilhard de Chardin (1959), for example, makes much of man's special place in the cosmos. He ingests the whole Darwinian story and then invests it with a mystical slant, claiming special significance for the phenomenon of man. Much of what we say subsequently in this book is concerned to locate man more dispassionately into the order of living things. We hope to restore theory to its proper place and not to inflate it as Teilhard did, into a new theology succoured on Darwinian scraps. Finally a word about latter-day religious opposition to Darwinian views. At present court hearings are taking place in America to decide whether biology school teachers should provide their pupils a wholesale genesis-based alternative to Darwinian theory. We shall have to make do with speaking past these fundamentalists; and shall try instead to offer the reader an integrated view of the position of man in the natural order.

As far as the 'missing link' is concerned, this phrase might now have become a misnomer. Recent discoveries by paleontologists have thrown up several contenders – Neanderthal man, Java man, Australopithecus and Homo habilis (Reader, 1981). It is probably more appropriate now to talk of the 'missing chain' – that is, an interposed line of fact and conjecture forging these links into an adequate description and explanation of our own line of descent. It may now be not so much a matter of gaps in the biological substratum that paves the way to man; but only a matter of the binding account of how man's special capacities – linguistic, moral and conceptual – could have evolved within the natural history of forbears whose fossil and artifact remains have already been discovered.

One of the first critics to cite the phenomenon of mind as an alleged obstacle to Darwinian theory was the Duke of Argyll in 1869. This objection would, no doubt, be felt much more keenly by those steeped in the then current philosophical and religious milieu. Cartesian dualism had taught that mind was wholly distinct from matter. It would not be surprising if the phenomenon of allegedly disjoint and transcendent mind posed less of a problem to the interdisciplinary intellect of today. Current observation and theory from ethology, cognitive psychology, neurophysiology, cybernetics and artificial intelligence are helping to clear out the last corners into which phenomenologists and dualists might try to retreat.

A close cousin to the missing link argument is the 'improbability of emergence' argument. This runs as follows. In the case of a complex organ like the eye it is sometimes hard to imagine how the many intermediate stages could have served a useful purpose, and have been selected for. The delicate componentry is highly interdependent: outside the context of the whole eye, for example, the retina would have no functional significance; likewise the cornea, the rods and cones, and the optic nerve cannot be accounted for except in conjunction with all the others, as they all go to make up a single organ with functional significance. One version of this argument goes further in maintaining that such wondrous integrated complexity would have arisen only by the most fortuitous ensemble of simultaneous mutations. These genetic saltations, as they are called, are then held to be so improbable that their relatively frequent and independent occurrence in different phyla must be evidence for divine intervention in the evolutionary process.

There are two things wrong with this argument. The first is that it neglects comparative anatomical studies of the eye from its most primitive form as a single light sensitive cell to its most developed form in higher animals. Moreover, Riedl (1977) has suggested ways how such interdependent complexities could have evolved without saltations. The second thing wrong with the argument is that it abuses or misuses the notion of probability. Torn from its usual context where it attaches to events in a limited domain against the background of established regularities the notion arguably ceases to apply to unique historical happenings such as mutations, mathematical insights and poetic inspiration. In the domain of human affairs it is far more appropriate to speak, for example, of probabilities of individual voting at the polls, than of declarations of independence or declarations of war. The two most important ways in which we understand probability are, firstly, long-run frequencies of repeatable events (such as coin tossing or nuclear decay) and, secondly, individual subjective betting rates (concerning matters such as the weather and horse races). It is no accident that modern probability theory was born at the junction of these two domains, namely gambling with cards and roulette wheels (see Hacking, 1975). Now, those who advance the argument from 'improbability of emergence' as it concerns either genetic saltation or accumulative mutation are arguing for an allegedly infinitesimal probability that *any events of this kind should have taken place at all*. But, we ask, who could place a bet on the question of whether the eye arose from any genetic mutations at all? Such a bet simply would not make sense. Moreover, who can make sense of the notion of the long-run frequency of such 'events' as *that there have been the appropriate genetic mutations at all*? There simply is no such long run delivering itself of the required frequencies.

This is not to say, however, that it would not make sense within a theory of actual and possible mutations to attach probabilities to such-and-such mutations occurring. For, the theory now concedes the existence of the sample space in question: and, if it can get clear about the nature and range of events in this sample space it can perhaps be used to determine probabilities as long-run frequencies of the same.

Imagine 52 draws of single cards from a pack containing 52 cards. The probability of any completed series is one in 52! If we increase the number of available cards and/or the number of draws in a series, then the probability of any one series vanishes at an astronomical rate. If, however, Methuselah dealt himself a pack of 1,000 cards, would we deny him his game of solitaire on the grounds that the outcome that he held in his hand was ridiculously improbable? Such is the force revealed in the fallacious argument from 'improbability of emergence'.

Suppose that at some future time one of a set of possible outcomes is to be realised. If we calculate beforehand that one of them has a very low probability, and then that very one is realised, this will be cause for surprise. If someone repeatedly predicted such improbable outcomes, we might wonder whether he had divining powers or access to a theory that drastically redistributed the probabilities in question. But when looking at evolution, we are looking into a past in which no such predictions could sensibly have been made. That what happened did happen cannot be any cause for bewilderment. At any point in time there was a large number of possible future courses. One of these was to be taken. The probability that some one would be taken is 1. One must also be mindful of the extent to which selection can change probability distributions. Suppose some evolutionary change requires three specific mutations, each with a probability of 10^{-5}. Then the probability that all three occur in one germ cell is 10^{-15} – a very small number compared to population sizes, which range between 10^5 and 10^{12}. This probability can increase if reproduction is sexual, for then the three mutations do not have to occur within the same germ cell, or at least germ line, in order to be expressed together in any organism. But natural selection can drastically change the picture. Suppose each mutation brings selective advantage on its own. After the first occurs it will spread through the population. Thereafter the next would do so, and the third likewise. Thus each would be much likelier to find the others' company in some organism or other: a threefold event with initial probability of 10^{-15} would, within a few generations, have come to have the probability of 10^{-5}. In a population of 10^{10} individuals this is near certainty.

So far we have dealt with criticisms of evolutionary theory that we consider to be relatively easy to counter, criticisms inspired in the main by religious beliefs about supernatural agency and the uniqueness of man. There are more substantive, mainly recent criticisms to be made within the general scientific framework, to which we shall address ourselves below.

Our immediate task in the next section is to fill in some of the necessary detail from genetics and basic biology to put our foregoing discussion on a firmer footing. The reader familiar with the most basic notions of modern biology may omit the following section. In section 18, with the biological details behind us, we shall return to the discussion of the scientific status of evolutionary theory. In the intervening sections we shall be concerned not only to map out a coherent stretch of scientific 'knowledge', but also to clarify certain key concepts of biological theory, and explain current technical terminology.

2 ALLELES, GENES, CHROMOSOMES: THE BEARERS OF GENETIC INFORMATION

All higher animals, including man, have in each of their many billion cells a cell nucleus. This contains a set of *chromosomes* whose number is characteristic of the species. Chromosomes occur in homologous pairs in each nucleus. Human beings have 46 chromosomes in each nucleus, arranged in 23 pairs. In the formation of a germ cell, that is a sperm or an egg, one member of each pair of chromosomes is present in the newly formed nucleus. Thus the germ cell contains 23 chromosomes. When a sperm unites with an egg upon fertilisation, the newly formed *zygote* once again contains the characteristic number of chromosome pairs. In each of these pairs one member comes from the father, the other from the mother. Successive cell divisions starting with the zygote eventually produce a new individual. Exactly how this proceeds, with cells arranging themselves in patterns and becoming ever more specialised in their function, is one of the outstanding problems exercising biologists today. Chromosomes can be seen with a light-microscope and reveal themselves as longish thin forms. They consist of interlocking molecules of protein and Desoxyribonucleic acid, the latter well known by its acronym DNA. DNA consists of sugar molecules, phosphoric acid and four different bases arranged in the double helix structure first discovered in 1953 by Watson and Crick. This discovery yielded the key to the problem of heredity. Its gross chemical and structural properties allow DNA to make copies of itself (to replicate) and also, through the mediation of so called RNA molecules, to provide the crucial information needed for the synthesis of proteins. Proteins are the major functional and structural components of living beings. They are made of chains of amino acids, of which there are about twenty. Each of these twenty amino acids is coded for in the DNA by a triplet of nucleic bases. Thus a sequence of such triplets determines the sequence of amino acids of any protein produced in the cell. The sequence of bases that codes for one protein is called a *gene*. Chromosomes come in different sizes, with an average human chromosome containing many thousands of genes. It is not known how many genes code for the structural complexities of a human being or any other vertebrate animal. We have enough DNA for about 1,000,000 average sized genes. But we know that much of this does not code for proteins. So if it has any function at all, it must be to regulate the coding DNA or to structure the chromosome. (For details of the exciting discoveries in molecular biology in the last decade, for example, Alberts et al. 1983).

At this point we must distinguish between two senses of the term 'gene'. In the first sense, genes are macromolecules of a particular chemical composition. In the second sense, a gene is identified as that type of macromolecule whose informational content determines a particular protein. Two genes may be distinct in the first, chemical sense, but identical in the second, informational sense. This is so because with four units of information (the four bases in the DNA) one can form 64 triplets (4^3): but there are only about twenty amino acids. Thus an amino acid can be coded for by more than one triplet. Chemically distinct triplets may therefore be informationally synonymous 'words' of the genetic code. Whole genes, being composed of triplets, are like sentences in this code; thus chemically distinct genes may also be synonymous as 'sentences'. Later on when we discuss gene selection we shall be talking of genes in the second, informational sense; and what we shall have to say will depend on bearing this distinction in mind.

The base sequence of a gene can be changed, especially during replication, when errors can occur. Then strictly speaking the gene would not be replicating itself but would be producing a chemically and possibly also informationally distinct version of itself. Changes in the informational content of a gene can also result from chemical events induced by radiation and other as yet unknown causes. Such changes are called *mutations*.

On a chromosome the genes occupy different *loci* (positions) and may be thought of as beads on a string. Each locus is associated with the production of a protein that is to serve a specific function within the individual. Enzymes catalyse various chemical reactions, other proteins are involved in the construction of cell walls and so on. As we have already indicated, mutations can supply new versions of a given gene, with each version able to serve at that locus more or less satisfactorily. We say 'more or less' because some mutations may turn out to be deleterious, while yet others may prove to be advantageous in comparison to the existing ones. The distinct versions available for a given genetic locus are called *alleles*. The *gene pool* of a species is composed of all extant alleles for each genetic locus in the inventory of chromosomes.

Any species which bears its chromosomes in pairs as we do is called *diploid*. A locus, as it were, cuts across both chromosomes of a pair. If the two alleles at those respective points are distinct then we say that the individual is *heterozygous* at that locus. If, however, those two alleles are identical, we say it is *homozygous* at that locus. In a heterozygous individual it can happen that only one allele expresses its information, in which case it is said to be *dominant*, while its unexpressed partner at that locus on the paired chromosome is said to be *recessive*. These notions, of course, apply to alleles taken two at a time. An allele may be dominant with one partner and recessive with another. When both alleles are expressed we have a case of intermediate inheritance. An example of intermediate inheritance is the skin colour of children of mixed African and European parentage – although, to be sure, this is a somewhat infelicitous example in so far as skin colour is determined by the action of many genes at different loci. Some cases of intermediate inheritance of traits governed by single loci are known, such as flower pigments. But they are generally more difficult to find. We shall be devoting a whole section to the distinction between such *monogenic* traits and traits that are *polygenic* in that they depend on the action of more than one gene.

The laws describing the distribution of alleles across descendants were discovered by the Augustinian monk Gregor Mendel, and published six years after Darwin's *Origin of Species*. Every school child today has heard of the three Mendelian laws of inheritance, but Darwin himself never came to know Mendel's work. It is an intellectual tragedy that he reached the end of his life without learning of these laws, as he knew full well that ignorance of the mechanism of inheritance was the crucial lacuna of his theory. At one point Darwin toyed with the notion of Lamarckian inheritance, which we shall discuss more fully below. He also considered blending inheritance, by which the hereditary material of the parents would be mixed in their offspring in a fashion analogous to the mixing of two pigments on an artist's palette. He failed, as others have done since, to find evidence for the former, and clearly saw the abovementioned limitations of the latter.

Mendel's discovery was first corroborated at the beginning of this century and only since then have his laws become part of general biological knowledge. The biologists Fisher, Wright and Haldane successfully incorporated Mendelian genetics into Darwinian evolutionary theory, thereby creating what is known today as the synthetic theory of evolution (Dobzhansky *et al.*, 1977). This has now attained the status of a paradigm, in Kuhn's (1970) sense of the word, and is the theory with which we shall seek to make our philosophy continuous. We shall, however, be extending and refining this theory through conceptual reanalysis (the philosophical contribution) and by attending to the latest research developments in the biological sciences (the empirical contribution). The latest fruit of the synthetic theory is sociobiology, with whose implications we shall be concerned in a following chapter.

Evolution is possible only when variation is at hand on which selection can get a grip. Such variation is attributable to two main sources. The first, mutation, we have seen to be the key source of primary variation. The second, *recombination*, occurs in sexually reproducing organisms in a way we shall presently explain. It ensures that single mutations arising in different individuals can be brought together in ever new combinations in new individuals. We described earlier how chromosome pairs split, yielding one member to a newly formed germ cell. When recombination takes place this is no longer strictly accurate. What happens then is that different sections of chromosomes from *both* members of the pair are recombined to form the single chromosome that is to be contributed to the new germ cell.

This vastly increases the combinatorial possibilities of bringing new mutations together on the same chromosome in a new individual. Recombination thereby amplifies the range of genotypes that are created in the future.

It is important that both mutations and recombinations are causally completely independent of their selective values. This is the sense in which one speaks so emphatically of *random* mutations and recombinations as providing the raw material for natural selection. If certain environmental changes were to confer a particular advantage on a theoretically possible mutant or recombinant genotype not presently represented in the gene pool, this would neverthless fail to make any more probable the occurrence of such mutations or recombinations as would produce such a genotype. We may now define the notion of a genotype more precisely than we could in the first section. The genotype of an individual is the profile of alleles (in the informational sense) throughout its chromosomal complement. Evolution in the population genetic sense (micro-evolution) may now be defined as change in the frequencies of different alleles for each locus in the population of genotypes.

3 FITNESS

Random and undirected genetic change cannot result in adaptive evolution; differential reproduction leading to adaptive change cannot be a matter of chance or randomness. This is not to say, however, that all fixed traits in actual populations have selective significance. Indeed, as we shall see in sections 9 and 10, one school of evolutionists holds that not every fixed trait in a population owes its fixity to its adaptive value. Chance might have had a large say in the history of life.

The theory of selection requires that in a constant environment certain alleles will proliferate rather than others, with an asymptotic approach to a state of optimal adaptation. Newly arising mutations can, of course, disturb the form of the approach as well as its asymptote. In moths not preyed upon by birds, the frequency of alleles for dark colouring could increase in one generation and decrease in the next, and generally vary quite arbitrarily. For genuine evolution it must be the case that some moths have more offspring *because* they have particular properties ensuring high reproduction. And, as we saw, this was indeed the case when birds preyed upon them and smog darkened their resting places.

The biological notion of fitness is a key concept of Darwinism in explanations of differential reproduction. In the modern theory of population genetics, fitness is a measure of the *relative* number of descendants that a particular allele *potentially* enables an *average* bearer to have – all this being understood, moreover, with respect to some fixed environment. We now briefly explain the three emphases in this definition, in reverse order.

Average because the success of an allele depends on the combination of alleles at all other loci in the individual in question. Placed in one genotype an allele might work to great advantage, but placed in another it could do quite the opposite. *Potentially* because if two individuals are genetically identical, but one is squashed by a meteorite before coming of reproductive age, we would nevertheless wish to ascribe the same fitness to them even though the latter failed to produce any descendants. The meteorite is a mishap which simply cannot be correlated with any capacity for survival of the stricken individual. But by contrast, in the case of a car accident one may not be justified in a similar claim, since many individual traits – ability to concentrate, foresight, keen vision, steady nerves, quick thinking, fast reactions and brute strength – could all help one to avoid or survive an accident.

Finally, in population genetics fitness is always measured as a *relative* magnitude. The allele that would on average – that is, after being considered in all genotypes – enable its bearer to have the most descendants, is ascribed the fitness value of 1 and is taken as a reference case. Any other allele at this locus consequently has a value between zero and 1, determined as the ratio of the average number of its offspring to that in the reference case. Naturally only alleles at one and the same genetic locus enter into such comparison when determining fitness values. It makes no sense at all to speak of the fitness of a gene at one locus in comparison with that of a gene at any other locus, since two such different genes can never enter into competition with one another. Alleles do, however, compete with others *at one and the same locus*, and those with higher fitness values will displace the others from the genepool. We have thus explicated the notion of the fitness of an allele in a given external environment. Note that throughout the explication the environment was assumed to be fixed. Obviously the distribution of fitness values to alleles at a given locus can vary – perhaps quite drastically – as the environment varies.

It is often alleged that the concept of fitness has only tautological uses. The question as to which is the fittest member of a given class has been answered by looking to see which of them has the greatest number of descendants. Then when questioned as to which member has most descendants the answer is supposedly that according to the theory the fittest one should. The useless circularity thus revealed allegedly empties the theoretical claim of empirical content.

A conceptual correction is called for here. The *unit of selection* and thereby the bearer of the property fitness is not the individual but the allele. This will be proved in the next section. Although it is definitely alleles and not individuals to which it is appropriate to ascribe a fitness value, nevertheless the fitness of an allele can only by measured by reference to individuals containing them. So the common practice is to ascribe fitness values only to individuals – even though this is theoretically incorrect.

In much the same way Mach's principle in physics has it that the mass of a body is a property conferred upon it only by reference to all the other bodies in the given inertial system. Analogous also is the Fregean principle in the philosophy of language (Frege, 1879) that the meaning of an individual word derives from its systematic contribution to the truth conditions of whole sentences in which it occurs. No doubt the reader can bring to mind further examples of holistic dependencies of this sort. In the case of the fitness of an allele no single measurement of numbers of descendants is guaranteed to yield the proper value, for the number of descendants depends importantly on two further factors. The environment can (as in the case of the meteorite) reduce or extinguish potential reproduction of an otherwise promising allele; or, on the genetic side, pleiotropy and epistasis (to be discussed later) may limit or increase its performance in a singular fashion. For these reasons one has to resort to measurements involving different individuals carrying the allele in question (thereby averaging out the effects of different genetic backgrounds) and also place the same in several different environments that are none the less typical for the species (thereby averaging out for that dimension also).

Once measurements have been made in this way then one is in a position to predict not only which allele will increase at the expense of others but also to make a probability estimate of the number of descendants to be expected from any individual whose genotype one knows allele by allele. But of course it is almost always practically impossible to do this for organisms in their natural state. The accusation of circularity is nevertheless in principle defused. We emphasise the words 'probability estimate' in the latter connection since of course it is a logical possibility that each fittest allele in the company of all other such alleles at other loci is disadvantageous: that is, ironically, that the respective fittest alleles make one another undesirable company (from the point of view of their bearers). This is, however, a very remote possibility, since it is likely that the particular 'chunk' of undesirable company for each allele would often have been present in other genotypes as well, thereby reducing its measured fitnesss value. Indeed the logical possibility we have in mind here is as remote as the hackneyed example from statistical thermodynamics of a table suddenly rising into the air as a result of fortuitously simultaneous bombardment by air molecules from below. In principle if one knows the biochemistry of an allele's genetic products one ought to be able to estimate the fitness of alleles that are theoretically possible, but absent from the gene pool. It was only because of this that Eigen (1971) was able to introduce the notion of fitness unproblematically into his equations. The two long papers in which he did so are a monumental attempt to unite the synthetic theory with thermodynamics. In prebiotic evolution one is concerned only with self-replicating molecules without organismic carriers. There, fitness is nothing but a certain algebraic function of survival rates, biochemical efficiency and rates of reproduction of the molecule concerned. In a system such as an organism that consists of countless molecules mutually influencing one another, it becomes impossible on purely practical grounds to calculate fitness from the molecular parameters. One has to resort in this case to global system parameters and make do with a crude estimate of fitness values. These values are inaccessible in the wild, and satisfactory measurements are only ever achieved under laboratory conditions contrived by population geneticists. But this provides no ground for regarding fitness as a useless or inconsequential measure, thereby bringing the whole theory of evolution into discredit. Even though some biologists consider the practical inaccessibility of fitness values as a serious defect of this theory, their eschewal of it and cognate notions is an overly positivistic response to experimental limitations. The situation is not without precedent in other areas of perfectly respectable science, such as quantum mechanics and relativistic cosmology, both of which deal with theoretical magnitudes which practically elude measurement.

Sexual behaviour, an essential part of the total behavioural repertoire of every higher organism, is often engaged in to assess the fitness of potential mates. Male fruit flies dance, sing and probe the females with their feet and tongues. While they do so, they are scrutinised by the females as though being auditioned for the role of reproductive partner. To a lesser extent the male avails himself of the opportunity for similar appraisal. (For the exact reason why the male is less critical in his choice of a sexual partner, see the section on sexual selection.) What concerns us here is the fact that the *Drosophila* females are behaving in a way best explained by describing it as her estimation of his fitness (see Ehrman, 1972, for a review). So perhaps the *Drosophila* female, and females of many other species besides (see Blum and Blum 1979 for insects), including our own (Cooke and McHenry 1978), have a shrewder heuristic for determining fitness values or at least effecting fitness comparisons than do population geneticists burdened by combinatorial detail!

Thus far we have concerned ourselves exclusively with the fitness concept of the population geneticist. There is a similar concept in evolutionary biology that attaches to populations or species rather than individuals. These two concepts associated with the word 'fitness' can coincide in particular domains but are sufficiently distinct to confuse the thinker who conflates them. Not only individuals compete constantly for scarce resources. Very often different species compete for an ecological niche that only one can win. In such a case the bearers of fitness (although, as we shall see, not the ultimate ones) are not the individuals but are rather whole species. With individuals it is already difficult enough to specify what features, apart from reproductive success, are to be measured when determining fitness. This, as we have seen, threatens to taint the process with undesirable circularity. Measurements of fitness are even less feasible as far as whole species are concerned. Population, size, 'intrinsic capacity for increase', and the population's competitive ability have all been used and have been found deficient in one way or another.

At the individual level, for example, it is relatively unproblematic to specify a time at which the number of descendants is to be counted. The descendants in the first generation present no complication except for such exotic mutations as 'grandchildless' in *Drosophila* which allow children to be produced but not grandchildren. At the population level it is insignificant whether the near future will see more or fewer members than at present. Indeed, short-term population increase could severely overload the resource base and lead to population crashes from which the species might not recover; this would surely provide ground for ascribing the species a low fitness value. The only interesting question is whether present degrees of adaptation will secure its survival in the distant future or whether the species will then be extinct, like almost all of those that have ever roamed the earth. How far ahead this is to be determined – whether 10^8 years as Thoday (1953) suggests or only 10^3 years – is hardly important. All that matters is to descry properties that can be measured in different species, which secure their long-term survival on our planet. A decisive factor ought certainly to be how independent they are from the environment. The latter can change in unpredictable ways, and organisms that cannot adapt quickly enough will die out as soon as the going gets tough. An original suggestion in this regard has come from Bruce Wallace (1977). He maintained that a species' ability to control its own population size is of the utmost importance. A species that could regulate its numbers in response to resource levels would avoid the population crashes we spoke of above. Methods of such regulation include territoriality (as emphasised by Wynne-Edwards, 1962), which ensures that areas do not become too densely populated, or processes that amplify available genetic variability in response to environmental influences. A case of the latter is preferential feeding of the most robust descendants at the expense of weaker siblings. It actually occurs in many species of birds and mammals.

If one could say how the environment will change then one could calculate particular species' survival chances by using a composite of such measurements. Since we cannot do this we must make do with general expectations which will no doubt disappoint those (like J. Maynard Smith, 1977a) who would like to see evolutionary theory make predictions as powerful as some physical theories. This, however, does not in any way mean that evolutionary theory can make no predictions at all and therefore forfeits its scientific status. But more of this in section 18 below.

4 THE UNIT OF SELECTION

A certain form of behaviour is advantageous to a species and has therefore been favoured by natural selection. Evolutionary explanations along these lines have been given even until quite recently by competent biologists. Konrad Lorenz (1966), for example, held the view that combative inhibitions and forms of submission in species such as wolves or red deer, which are well equipped otherwise to maim or kill their fellows, have been selected because they serve the interests of the species rather than those of individuals or alleles. But much better explanations of aspects of social behaviour such as combative inhibition are to be had from the theory that ascribes a predominating causal effect to the underlying ensemble of genes. The *locus classicus* in this connection is Williams's *Adaptation and Natural Selection*, in which he convincingly shows how far the genetic story accounts for the characters and their characteristic social plots. The book was a critique of the earlier group selectionist account proposed by Wynne-Edwards (1962). The latter's book *Animal Dispersion in Relation to Social Behaviour* contains a fund of observations of what he called epideictic behaviour, forms of behaviour enabling members of a species to make a population census in order to adjust their rate of reproduction to the resource base. The theory of gene (or better, allele) selection, however, which originated with early population genetic theory and was consolidated in Williams's manifesto, reveals just how implausible such group selectionist accounts are. For behaviour interpreted in Wynne-Edwards's way is not what Maynard Smith (Maynard Smith and Price, 1973) would call an *evolutionarily stable strategy*. A genetic mutation for selfish individual profit from breeding restraint by conspecifics would rapidly undermine group solidarity in this respect.

In individual selection and group selection different individuals or groups are supposed to be competing for reproductive space. Now on the lowest level it is alleles and not genes that compete with each other. Thus by analogy the force of selection should be called allele selection. But this is a semantic point of perhaps little import. Much more important is the error of compounding the unit of selection with the force of selection. Natural selection, sexual selection, individual, group and artificial selection are all forms of selection specific to certain environments. But all these forces work on the same unit – namely the allele.

Gene, individual and group selection are widely taken to be three alternative forms of selection, each with its own unit. But our own view of the matter is that there is only one unit of selection, and that is the allele. All adaptive evolutionary change is based on change in allele frequencies. Alleles are informational units expressing themselves phenotypically on many levels of differing complexity: from DNA-structure to subcellular and cellular structures and from there to organs, individuals, groups (kin- or otherwise), species, genera, families etc. If members of any given level can be seen as in competition with each other for resources then they can be ascribed fitness values, in the sense that they critically express the underlying allelic differences for the selective vectors. For phenotypic differences that reflect genotypic differences between members of one level will be subject to various selective forces; and adaptive evolutionary change results if these forces cause shifts in the allelic composition of the genepool. Thus individuals, groups and species can be fit or unfit in a population genetic sense, whereas organs and subcellular structures cannot.

Each of these organisational levels is shaped by its own selective forces. Forces on one level can conflict with forces on another. These cases, naturally, are of special interest, and have been given special names. Individual selection, group selection and sexual selection are just such forces, acting on different levels and sometimes in direct opposition to each other in that they force allele frequency shifts in opposite directions. Group selection can result in individual altruism. Altruists may breed less and so help to ensure group survival. This might, of course, in the long term guarantee individual survival; but in the short term individuals stand to lose if they exercise restraint in breeding, for this allows less altruistic members of the group proportionately greater representation in subsequent generations.

When talking of individuals' interests we should not forget that the individual is no more than a particular environment which alleles have shaped for themselves, and within which their project of replication is executed. Also, group life is just one particular environment that gives rise to new selective priorities for the genes that have made it possible. These environments are largely created by the action of genes. But for a given allele the most important feature of the environment determining its future success is all the alleles on the other loci in the individual that they are together helping to shape. Similarly, if a certain part of the genepool of a given species makes group life possible and advantageous, then other parts will prosper in so far as they collaborate in that project. Again, when sexual reproduction arises and the two sexes make different parental investments, then a part of the genepool will be confronted with a new environment – one making sexual advertisement, intrasexual combat for mates and so on possible strategies for higher replication.

All these selective forces exert their influence, however, through changes in the allelic composition of the species. It is therefore incorrect to think of individuals and groups as alternative units of selection. Alleles are the units whose frequencies are shaped by various selective vectors some of which have been given special names, such as group, individual, sexual and artificial selection, because of their special interest or importance. For example, group selection is the selective force exerted on alleles as a result of the fact that the individuals carrying these alleles live in groups. Individual selection is the force acting on alleles as a result of the fact that they are located in the cells of individuals. Free alleles would not be affected by epistasis (influences exerted by alleles at other loci). Gene selection, it would seem to follow, is the pressure exerted on alleles as a result of their pairwise existence in diploid organisms. (Selection for dominance, for example, makes no sense for haploid organisms.) These forces are just components of the environment an animal lives in. If talking of group selection is admissible, so too can we speak of altitude, temperature or humidity selection. The group is an environmental parameter for the alleles, for their frequencies are affected by the contingencies of group life. But other contingencies affect those frequencies – such as humidity, temperature, etc. If two selective forces oppose each other then allele frequencies will change in the direction of the stronger force or not at all. Change in allele frequencies is thus analogous to the movement of a physical body, under the influence of physical forces.

We believe one should reserve the label 'unit' for the entities ultimately responsible for adaptive evolutionary change. Other entities might also compete with each other – individuals and groups might do so, for example. But all adaptation resulting from such competition is due ultimately to allelic reshuffling. Individuals and groups are interesting units only in so far, as Hull (1981) points out, as they are viewed as *interacting* with their environments rather than as *replicating* within them. Interactors need not be replicators. Replicators (alleles) count as units of selection in the above sense because their structure is stably preserved through many events of replication. The same is not true of individuals. Individuality resides in a unique combination of alleles. In the case of higher organisms incapable of asexual reproduction, no genotype is replicated reliably enough to guarantee its cohesion over successive generations. Reliability of replication is what is most importantly at issue in the identification of units of selection. The only things that replicate reliably are alleles (and possibly also memes – see Chapter 2). Through successive accurate replications, allele identity is conserved over many generations. The mutations which do occasionally modify them occur at a low frequency. Because germ cells are haploid and recombinations occur within chromosomes, genotypic cohesion is orders of magnitude lower than genic cohesion.

The concept of gene developed here for evolutionary biology is different from that of molecular biology. Closely linked loci on a chromosome cannot recombine freely. Intensive selection on one locus always affects allele frequencies at neighbouring loci. This is the so-called 'hitch hiking effect': a slightly deleterious allele can be carried along by a more advantageous neighbour, even to the point of displacing fitter competitors for the same locus. Furthermore interactions among genes may also cause 'linkage disequilibrium': certain alleles on one locus cease to be randomly associated with other alleles on the same chromosome (Franklin and Lewontin, 1970). In many organisms whole groups of loci are protected against recombination and are always inherited as a chunk. Breaking these up adversely affects individual fitness. It is becoming clear that the genome is not as static as had been imagined in the early days of genetics. Genes do 'jump' between loci and produce more copies under stress (Lewin, 1981). Dawkins (1976) realised that this meant that the allele of the molecular or of the population geneticist need not be the unit of selection. It might often be a whole chunk of a chromosome, consisting of several loci, that forms a unit for selection. But this discrepancy does not at all affect our discussion above.

In the following discussion one has to bear in mind that group selection is a selective force rather than a selective unit. Groups that change the selective conditions for alleles hitherto housed in asocial bodies can be nuclear families, wider kin groups or communities of individuals of arbitrary origin. In the human case tribes, city states and nations could count as groups for such purposes. What confers identity upon a group is, of course, a complicated matter, and an individual could very well be taken as belonging to different groups for different explanatory purposes. It is not necessary that members should live in close proximity or display any readily discernible cohesion. If they communicate and collaborate well enough then group members can pursue joint purposes, these purposes indeed being an important part of what makes them members of one and the same group, even though they be flung far and wide. For example, elites of all kinds typically display bonds, ties, allegiances and reproductive arrangements that cross normal frontiers and make them more or less independent of locale and neighbourhood.

Group selection can reinforce the operation of individual selection or it can counteract it. The latter is naturally the most interesting case, and concentration on it has perhaps encouraged the mistake of thinking that in the correct selective story one and only one of these forces may be regarded as being at work. In the more catholic account that we would urge, however, one would have to tackle the admittedly difficult task of apportioning the credit for evolutionary change where it is due.

One of the areas in which individual selection and group selection are sometimes seen as opposed forces is altruistic behaviour in socially living species. In addressing themselves to this problem area in particular, sociobiologists have advanced the allele selection paradigm tenaciously. They have broadly succeeded in entrenching Williams's principle that enjoins biologists always to look to what we would call allele selection for the fullest account possible before resorting to the supposedly vaguer, less tractable notions of group selection. The sociobiologists present the following alternatives. (1) Altruistic behaviour arises by group selection. (2) It is due to kin selection. (3) It is altruism which, if directed outside the kin group (thereby rendering it inexplicable by kin selection alone) can nevertheless be accounted for as so-called reciprocal altruism, which in turn can still be explained within the framework of allele selection. The first case is pretty summarily dismissed, or at least ignored as a matter of methodology by modern sociobiology. The other two cases and the terms 'kin selection' and 'reciprocal altruism' mentioned therein must await fuller discussion in a later chapter wholly devoted to sociobiology and its implications for our understanding of man's social and moral nature. Suffice it here to observe that group selection has not proved a very fertile source of biological explanation. At present all we have is a general definition of what it would involve were it to take place.

There is, however, one modern author in particular, namely D. S. Wilson (1980) (not to be confused with E. O. Wilson), who has been concerned to rehabilitate group selectionism after the heavy criticism it sustained from Williams and his successors. But his supposedly novel group selectionist model is no more than a highly general form of kin selectionism (see Maynard Smith, 1976a, Dawkins, 1979). The real merit of D. S. Wilson's case lies in the way it has caused gene – or kin selectionism – to flex its intellectual muscle and extend its grasp even further.

D. S. Wilson's models are intended to apply to many species of plants and animals whose breeding and dispersal patterns meet certain broad conditions that, he contends, obtain quite often in the wild. The groups concerned (which are all spatiotemporally concentrated) undergo cyclical disintegration and regrouping. Suppose that the species members live in groups occupying and defending their own territories. Suppose further that some groups include altruists who, in the group's interest, sacrifice reproductive opportunities in order to help their group as a whole achieve a higher rate of reproduction (on average). Normally, if there were egoists acting to their own advantage within such groups they would pose a threat to overall group survival – because, say, of precarious resources – and presumably the altruists would share in the group's demise. If, however, the groups undergo cyclical dissolution and are later reconstituted in a random fashion then the altruists can steadily increase their numbers. Groups including them can on average be more successful and the proportion of altruists within the species can accordingly increase. A precondition, of course, for such a course of events is that individual lifespans exceed those of the groups.

Further commentary on Wilson's model is appropriate here. Upon random reconstitution of groups a certain genetic type may come into contact with 'birds of a feather' more frequently than with other types. Consequently a basis can be established for mutual aid which extends not only to relatives (kin) but also conspecifics with partially similar genetic constitutions. But here is the Achilles heel in D. S. Wilson's claim to be on to something new. The gene selectionist need only point out that kin selection depends not on a *genealogical* notion of kin, but only on a synchronic and ahistorical one. Genealogically unrelated individuals can nevertheless (as far as the genetic theory is concerned) turn out to be kith and kin, simply by sharing a significant number of alleles. This is all that matters from the alleles' point of view. If they rig the bodies carrying them to recognise other bodies that do so, and to act accordingly in propagating their interests, then it would matter not a jot if kindred bodies were not genuinely – that is genealogically – kin. (There is experimental evidence involving frogs (Blaustein and O'Hara, 1981) demonstrating allelic recognition. The experiments were carefully designed to prevent the possibility that kin recognition might have been the result of mutual contact in early life.)

Despite this subsumption of D. S. Wilson's group selectionism by the wider embrace of kin selectionism there do, nevertheless, appear to be other cases of properties that could only have arisen by group selection. Lewontin (1970) recounts two. The first is the myxoma virus in Australian rabbits (or, rather, rabbits brought to Australia by European settlers). This was introduced in an attempt to kill the rabbits off, but what was not realised was that the virus was carried from one host to another by a fly that bites only live rabbits. The more virulent (that is, individually fit) the virus, the quicker it would kill its host. Flies would then not bite the corpse, so those viruses it harboured would reach a dead end. Less virulent strains that infected their hosts but let them limp along would be carried further afield by the fly. Consider now an arbitrary virus. If it is in a group of highly virulent fellows it is unfit; in the company of less virulent fellows, however, it is fitter. The overall virulence of the group determines a component of the fitness of viral alleles; thus group selection is at work.

At first sight this analysis is in order; but there is an individual selectionist case to be made quite generally for parasite-host relationships and selection for non-fatal virulence. Various factors deserve careful consideration, such as the relative proportion of a host's parasites that can be passed on to a new host, the possibility of reinfestation by unrelated parasites, and the possibility that the parasite may – as in the case of the rabies virus – so influence the behaviour of its host as to maximise its chances of spreading despite its high virulence. As is well known rabid foxes attack and bite anything that moves around them, thereby raising the virus' chance of host hopping.

Lewontin's second example is the t-allele of the house mouse. This allele is fatal in homozygous condition but in heterozygotes so affects sperm production, that, despite heterozygosity, 95 per cent of all sperms come away with the t-allele. For this reason it is called a segregation distorter gene. In mouse populations the frequency of individuals heterozygous at the t-allele locus is much smaller than would be expected from its segregation distorter action and homozygous lethality. One way to explain why this is so is to invoke a group selective force working against the t-allele. House mice live in small groups that could meet the requirements for group selection. Group selection is taking place because an arbitrary individual is more or less fit according to whether he does or does not belong to a vulnerable group whose members bear the t-allele. Such groups, of course, have a much lower expectation of survival or increase in numbers. The crucial reason for this is their small size and the consequently higher chance that the t-allele might wipe out its reproductive potential altogether.

Group selective forces that do not oppose individual selection are naturally very difficult to distinguish from the latter. We believe, however, that in the evolution of our own species especially, group selection must have played a decisive part from a relatively recent time to the present. As our ancestors lived in family groups and gradually developed tools and culture, particularly gifted and inventive individuals could have raised considerably the reproductive chances of fellow group members. One has only to imagine, for example, the discovery of new weapons or agricultural methods or success in taming and domesticating useful animals. Such developments might occur solely because of the talents of one particular group member. It is only to be expected that culture – that storehouse of such inventions – greatly increases the enduring benefits to the group of especially inventive individuals long after they have died, and does so even if they leave behind no descendants endowed with the possibly rare genetic inheritance underlying the inventive talents in question.

5 POLYGENIC AND MONOGENIC CHARACTERS

So far the reader might have formed the impression that we have been assuming a simple 1:1 correspondence between genes and manifest properties of the organism, be they morphological such as hair colour, or behavioural such as altruistic disposition. But no complex property of any organism is due to the action of just one gene. Genes code for proteins by completely determining their sequences of amino acids. But all characteristics of an organism above the level of proteins generally arise from the interaction of several of the latter. Consequently they depend on several genes for their expression.

Notoriously, properties have eluded satisfactory analysis by philosophers. They are not as sharp edged and naturally given as one would like them to be. Wittgenstein's notion of 'family resemblances' (1953) is beautifully anticipated in Darwin's discussion of the term 'species'. Most Wittgensteinian philosophers have rested content with an understanding gleaned only from knowledge of how certain words characteristically used to express 'properties' behave in ordinary language. It is not our task here to dust out the conceptual attic of the working biologist. Philosophical problems will constantly lurk in the background during a discussion of the multifarious and extraordinarily complex 'properties' that will hold our attention. We shall also use the terms 'characteristic' and 'trait' as stylistic variants on 'property'. Such properties as a person's height, intelligence or sociability, to mention but a few properties of interest to the biologist and social scientist, are naturally determined in part by a multitude of genes. By 'determined by many genes' we do not only mean that many genes are involved in the ontogeny of a trait, but also that many genes may be responsible for differences among individuals with respect to the trait. We shall, however, show later that it is theoretically possible for the former to hold while all variation be determined by at most a few genes.

The rarity of characteristics whose variability in a population is governed by single genes was, of course, the reason for Darwin's failure in his search for the mechanism of heredity. Mendel was most fortunate in his choice of characteristics (size and bloom colour of garden peas) whose major variability was due to variation at a single genetic locus, and whose sets of genes lay on different chromosomes and were therefore distributed independently of each other. If two genes lie on one chromosome then usually they are both bequeathed if either is. The exception to this is when they are separated by recombination. This complication could have prevented Mendel from arriving at his account of particulate heredity.

Polygenic traits are those which depend for their expression on more than one gene. We repeat that this is not to say, however, that no single locus could be responsible for much – even all – of the variation in that trait. Normally, however, continuous variation results from alternatives at several loci among those collectively responsible for the development of the characteristic in question. A useful analogy is that of the three-legged table whose legs may be thought of as genetic loci, with alternative legs of different lengths corresponding to alleles, and the height and tilt of the table top representing the polygenic trait concerned. One can drastically affect the tilt by shortening just one leg; and in general a change of any leg's length varies the trait described.

Of the human traits mentioned above, intelligence is a prime example of one depending, no doubt, on the action of very many genes but also especially vulnerable to single gene mutation. This is so whether intelligence lies behind a person's performance in IQ tests and is only imperfectly disclosed by it, or cannot be taken as amounting to anything more than what such tests measure. There are many known alleles for many loci, the presence of any one of which severely retards intelligence. (See any textbook on human genetics, such as Stern, 1973.) Indeed, single gene mutations almost always produce pathological extremes of continuously varying traits. (Some of these, like the phenylketonuriac gene, do so because their devastating biochemical effects 'short-circuit' the usual attenuated pathways of cause and effect that link genes with behavioural or cognitive capacities.) What is meant by a 'continuous' trait here is, of course, that the frequency of measured values of the trait is a continous function of the measured values themselves. That is, frequencies of slightly different values are themselves only slightly different, if at all.

Quantitative or biometrical genetics is the science of continuously varying polygenic traits (or characteristics, or properties, as we have also called them). The genetics of polygenic traits is complicated by the fact that their expression is always influenced by the environment to a greater or lesser degree. Quantitative genetics is, consequently, importantly concerned with assessing the relative contribution, to the production of phenotypic variation, of heredity and environment. When a trait varies, one wishes to know whether this is due to genetic differences between individuals, or whether they vary in this respect as a result of having been in different environments. Obviously this question is most important for plant and animal breeders concerned to know whether their breeding strategies are likely to produce the desired results. The greater the hereditary component, the more effective their selective measures will be. In our context the results of quantitative genetics are interesting and important because of controversies surrounding potential application of its methods to our own species. Some might see genetic manipulation of intelligence and moral fibre as a live possibility posing grave ethical problems. But it is far likelier that utopian ends can be achieved by more level-headed and common sense social arrangements. We could all be more convivial and fulfilled without tinkering with our genes.

Population genetics has so far managed to provide mathematical models only for evolutionary processes involving at most two genes. Unfortunately these models are completely irrelevant for evolutionary processes governing complex characteristics of morphology and behaviour. This is because these complex traits involve more than two genes. It is reasonable to expect that, even if the exact details in the more complex cases elude computation, nevertheless the overall qualitative picture or form of solution in the complex cases will be like those of the simple ones. In a similar way, although we have analytical methods for solving the Newtonian equations of motion for two gravitationally interacting bodies at most, we can, in the absence of exact curves in the many body case, identify stable forms of solution such as that of planetary orbits in general.

To illustrate the complexity facing a geneticist, we need only consider the 100,000 genes or more that each of us bears. A part of this endowment is for fundamental processes in cellular physiology, and the rest must suffice for programming 10^{10} nerve cells and goodness knows how many behavioural routines governed by these cells. The number of genes is far too small to determine even a small part of this neuromotor complexity in a single 1:1 fashion. It follows from this that the build-up of structures underlying complex systems must be under hierarchical and modular control. Each gene must influence many traits (pleiotropy) and every trait of some complexity must be influenced by many genes (polygeny). This conception has proved its worth in numerous investigations in behavioural genetics (e.g. Fuller and Thompson, 1978).

The *Drosophila* fruit fly's sexual behaviour provides a striking example of the pervasive and multifarious causal powers of single genes. The male courts the female by dancing, singing with its wings and touching her. Almost every mutation that has been introduced into the gene pool of this organism impairs the reproductive success of the male, frequently by impairing his sexual behaviour.

It is only reasonable to suppose that this is because the proper functioning of every conceivable aspect of a male fly's behaviour is involved in what one might therefore perhaps a little misleadingly wish to specify more narrowly as sexual behaviour. If the male fly cannot see he will find it difficult to follow and touch the constantly moving female; if he has no wings he cannot sing; if his motor co-ordination is impaired he cannot dance. Thus the behaviour identified as sexual behaviour for certain of our explanatory purposes is subserved by all the mechanisms and modules underlying any kind of behaviour at all. Thus almost every gene makes a contribution to sexual behaviour. Even if we limit our interest to a narrower band of the behavioural spectrum, such as singing alone, we shall still find at least small quantitative behavioural changes resulting from any mutation affecting the physical components (such as nerves and muscles and wings) underlying that particular band of behaviour. Conversely, when mutations occur that appear to affect only the behaviour of the individual, one can be sure that a change has taken place at some physiological or morphological level as well.

Thus when a breeder decides to select for a trait in a certain direction he must be prepared to accept unforeseeable changes in other behavioural respects. Indeed, so-called correlated effects are well known to animal breeeders, and were much emphasised by Darwin in Chapter 1 of the *Origin of Species*. Since there is only a small probability that a complex system will be improved by arbitrary changes, what may be regarded by the breeder as a positive change is usually accompanied by negative effects by the lights of natural selection. Intensively bred animals usually perish when returned to their forefathers' natural habitats.

The leading exponent of sociobiology, E. O. Wilson, seems in his major work (1975) not to appreciate how highly integrated a genome is. He writes as though he believed single genes could determine the development of complex behavioural traits. This is unfortunate because his case is in no respect weakened by a full appreciation of epigenetic complexity. In his latest work (Lumsden and Wilson, 1981), he appears, however, to have avoided such oversimplification. Nevertheless he and his co-author still write as though behavioural mutants lend support to sociobiological theory. But what a mutant shows is that the behaviour in question can be wrecked by single locus mutation, not that its normal development depends on the action of but a single gene. They do *not* show that genes are correlated in a simple one-one fashion with behavioural traits. Nor do they show that behavioural variation in a natural population is genetic in origin. What they *do* show is that such variation potentially exists. But this is a trivial point. To establish it one need not go to the lengths of producing behavioural mutants.

6 HEREDITY AND ENVIRONMENT

To avoid misunderstandings in any discussion of heredity and environment one must clearly distinguish two senses of the word heritability. (Lehrman, 1970)

The first sense concerns the individual development (ontogenesis) of an organism. In the fertilised egg is a genetic programme guiding the individual's development into an adult organism. At every stage of ontogenesis certain genetic conditions have to be fulfilled for further development to be possible. In addition, however, each increment of progress requires an adequate environmental set-up for the genetic programme to be carried out. In an environment without genes there is no life; and genes in a vacuum would never be able to express the information they contain. These observations are so trivial that one should not have to make them; but, simple though they are, they are often ignored in conversations among non-biologists. Presumably their neglect of the obvious results from the fact that genetic programmes are not all equally rigid. Many programmes are very sensitive to environmental influences (open programmes) while yet others are not so (closed programmes). Whitethroats can be reared in isolation in sound-insulated cages but will still command the full range of their species' characteristic calls and songs. By contrast, the bullfinch must be exposed to adult songs during infancy in order to be in *bona fide* voice in maturity. If one rears a bullfinch among canaries then it will pick up canary songs. In between these two extremes is the chaffinch whose song development follows both open and closed programmes. Birds reared in sound-insulated isolation sing with the characteristic number of syllables of characteristic lengths, but their arrangement of these into cadences does not follow the typical pattern (reviewed in Eibl-Eibesfeldt, 1967).

The differences between open and closed programmes are basically only quantitative. They stem from varying contributions from the genetic and environmental sources respectively to the total information required for the development of a trait. One could arrest the development of the whitethroat's song by the drastic 'environmental' influence of removing its vocal chords. Likewise one can imagine a mutation that could alter the innate or acquired repertoire of the bullfinch.

Ultimately every characteristic can be manipulated both genetically and environmentally. Extreme environmental influences can alter or extinguish even the most closed of closed programmes. And wholesale mutation can alter or extinguish any open programme.

Environmental manipulation can have awful effects on the expression of a trait – such as sexual behaviour of the rhesus monkey. The Harlows and their colleagues (1962) performed a series of notorious 'mother surrogate' experiments. Rhesus babies were removed from their mothers at birth and provided with various mother substitutes. This was all in an attempt to discover the key maternal influences on normal infant development. One mother substitute was made from wire and had attached to it a nipple dispensing milk. Another was made from wire, but covered with cloth. Yet a third was covered with electrically warmed cloth. Some babies were even fortunate enough to get a warm cloth version with a nipple thrown in. The results of these cruel experiments, the general nature of which could have been predicted without resort to such cruelty, were nevertheless interesting in their details. In all cases of extreme deprivation, adult sexual behaviour was severely disturbed. The more serious and prolonged the deprivation, the less responsive they were to the 'therapy' of being returned to a normal social group. The indications were that sexual maturation in the rhesus monkey follows an open genetic programme whose crucial inputs must be warmth, nipples, milk and so on. Moreover, the experiments disclosed that certain of these inputs were more important than others. Similar examples of extreme environmental deprivation are those hapless adolescents discovered in dark attics, neglected and maltreated by their parents, and growing up with marked behavioural and intellectual disorders as a result (Clarke and Clarke, 1976).

We pass now to an oft-cited example of an extreme genetic influence in producing a phenotype well outside the normal range of variation. Phenylketonuria is a single gene defect afflicting about 1 in 10,000-20,000 children. These children are severely retarded in their mental and physical development. The syndrome results from metabolic malfunction of the amino acid phenylalanine. Because the particular enzyme coded for by the gene in question is absent, a certain biochemical reaction fails to be catalysed and a metabolic by-product accordingly accumulates with disastrous effects. Thus this sickness is one which in the classical sense is genetically determined. Nevertheless it can be almost completely prevented by keeping to a diet free of phenylalanine until the age of about fourteen. Thus we have, surprisingly, a genetically determined condition that responds magnificently to environmental manipulation. A phenylalanine-free diet, however, is a highly contrived one, and a child brought up on it can hardly be said to be growing up in an environment within the normal range. And it is stability of a trait across the *normal* range of environments that marks it as one that is heritable in the ontogenetic sense. In this sense of 'heritable', however, no characteristic can be completely heritable. One cannot over-emphasise that every developmental increment requires both genetic and environmental inputs.

Matters are quite different, however, with the second notion of *heritability*, the one deployed in quantitative genetics. Trait differences within the normal range between individuals can be entirely due to genetic factors; or at the other extreme, can be entirely due to environmental factors. For example, all phenotypic differences that may be evident between genetically identical individuals (such as monozygous twins) are obviously due to environmental differences. Conversely, variations between individuals growing up in environments that are alike in every detail, such as bacteria in petri dishes, are grounded in genetic differences between them. Quantitative genetics provides suitable methods for the analysis of breeding experiments. It equips us theoretically to determine, in the case of all measurable traits, the relative contributions of heredity and environment in the creation of differences between individuals. In practice, however, insuperable difficulties stand in the way of applying these methods to our own species with the thoroughness that can be achieved with laboratory populations of fruitflies, or with farm populations of domestic animals and plants.

For the sake of completeness we mention here that quantitative geneticists employ two notions of heritability. These are heritability in the so-called broad and narrow senses. Total genetic variance is the sum of additive variance and so-called dominant genetic variance. Heritability in the broad sense is the ratio of this total to the phenotypic variance. Heritability in the narrow sense is the ratio of just the additive variance to the same.

A proper intuitive grasp of these notions is only to be had from a thorough study of the mathematics of quantitative genetics (see the two classics Falconer, 1960, and Mather and Jinks, 1971). Additive variance is best thought of as the reflection of the breeding values of all members of a population with respect to the trait in question. The breeding value of an individual is a measure of faithful phenotypic reflection of its genotype in its offspring across all possible matings. It is easy to see how this reflection might *not* be faithful. A mother, say, may display an extreme of some phenotypic trait owing to homozygosity of recessive alleles in many loci important to the trait in question. Her offspring across all possible matings would have many dominant alleles at these loci from their fathers, and so would not faithfully reflect the maternal phenotype. And these offspring, quite unlike mother, could themselves produce offspring once more very like her. With intermediate inheritance, which results predominantly in additive variance, the phenotype of the offspring expresses its genotype much more reliably. Therefore additive variance is all that counts for selective programmes, and is the most important part of the genetic variance as far as agriculturalists and eugenicists are concerned.

No trait displays only one type of genetic variation; and a further complication that we have not considered is that effects among the genes (epistasis) can enter importantly into the mathematical picture.

In the light of this second (biometric) sense of 'heritable', let us consider once more our examples above of extreme environmental and genetic influences on individual development. The contrast between these two extremes is strictly relevant to 'heritability' only in the first (ontogenetic) sense. Yet many people misguidedly try to draw, from cases such as these, conclusions about the origin of phenotypic differences among individuals. We are at pains to point out how fallacious such inferences are. We hope thereby to counter exaggerated interpretations of experimental and empirical findings by extremists on both sides of the nature-nurture debate.

Mutations can change any characteristic, but this does not prove that differences with respect to it within the normal range of variation among individuals are of a genetic nature. Conversely, the extreme consequences that may be inflicted upon individuals by extreme environmental conditions do not confer greater importance upon the causal role of environmental factors in the development of 'normal' differences among members of the species. In the second (biometric) sense of 'heritable', Harlow's rhesus monkey experiments and the existence of phenylketonurics do not strictly bear on the nature-nurture question. The quantitative geneticist, when investigating how much variability in sexual behaviour is due to genetic differences, would not in his right mind contemplate performing such experiments. Nor would an Eysenck or Jensen, in wondering how much of the variability in human intelligence is due to genetic differences, in their right minds tour hospital wards to carry out IQ tests on phenylketonurics and the like. Crippling environments and unusual genetic mutants lie outside the scope of the quantitative geneticist's concerns. Deprived monkeys, speechless adolescents discovered in dark attics, phenylketonuric and mongoloid children – in other words, extremes that can easily be attributed to single environmental or single genetic causes – are not relevant when one is trying to elucidate the relative contribution of genes and milieu to normal phenotypic differences. The two concepts of heritability are elegantly brought together in the concept of the reaction norm of a genotype. Each individual is capable of a certain range of phenotypic reactions to varying environmental stimuli. The size of this range will differ from trait to trait – open programmes having larger ranges and closed programmes smaller ones. It can also differ between genotypes – some being more adaptable, and others less.

We wish now to enter some evolutionary considerations that will permit us to make some synthetic connections between the two senses of the word 'heritability' just explained. (See Schilcher, 1982, for further treatment.) High variability between individual phenotypes with respect to a trait of selective significance is *per definitionem* always due to varying environments. In a stable environment we could expect variability to be very low, and to be attributable to recurring mutations adversely affecting fitness. In constant conditions evolution would taper off and come to a standstill. A phenotype that is optimal in such conditions would swell its representation within the whole population. Moreover, the frequency of mutations, which can also be subject to evolution, would decline, as would the rate of recombination involving the governing genotype.

The naive panselectionist, not without reason, sees phenotypic variability as a response to a heterogeneous environment. This variability can arise because of genetic reasons or as a result of individual adaptation, as in the case of learning. Which of these two alternatives occurs will depend on whether the organism's sensory motor system is geared to differences from place to place, and whether it is long-lived enough to have to cope with differences from time to time. Short-lived organisms with little time for individual adaptation will tend to exhibit the 'strategy' of genetically determined variability. A good example here is the way some fruit fly species' chromosomal constitutions vary cyclically with the seasons, as shown in the beautiful experiments of Dobzhansky and his school (see Dobzhansky, 1970). Organisms whose lives span several seasons and which are highly mobile will, should their central nervous system permit, tend to follow the course of individual adaptation through open programmes. Differences between such individuals in many a behavioural characteristic will therefore be due in large measure to the environment. But even in these organisms there will be genetically caused differences, in behavioural respects that cannot be left to turn on the course of individual experience, be this either because the nervous system does not have the required learning capacity or because the characteristic is required to be very constant and dependable. We do not wish to give the impression here that organisms can pursue only one of these strategies. On the contrary, this is seldom the case; most follow a mixed strategy.

In general, closed programmes will lead to individual differences being genetically determined and open programmes to their being environmentally determined. This may sound trivial and is grist to the nativists' mill. But it is theoretically possible that open programmes can lead to genetically determined differences, as when there are genetic differences in learning capacity. Conversely, a closed programme can lead to non-genetically determined variability, as when the programme's activation depends on spontaneous activity of the nervous system. Jacobs (1981) has worked out an elegant formal model which reveals all the different relations the two factors can enter into. From an evolutionary standpoint these last two possibilities are not at all likely. The first, open programme with high heritability, is improbable because it is highly unlikely that so much 'genetic' response as is required for the formation of a closed programme would have been forthcoming had not the environment frequently presented the organisms with the stimuli required to trigger the programme, thereby excluding much non-adaptive environmental variability due to spontaneous activity of the nervous system. Of course, if the environment is changing rapidly, these stimuli might become more rare, and others might enter, leading to precisely the non-genetically determined variability envisaged above.

Genetic considerations enter at every level of life, including the psychological and social level. Opponents of this view often maintain, perhaps quite correctly, that biological considerations cannot fully explain human behaviour. But extreme opponents are certainly wrong when they maintain that these considerations are utterly irrelevant or misplaced. Their proffered reason is that (allegedly) there is no genetic variability in human forms of behaviour. This, of course, is the key empirical issue on which sociobiologists take their stand, and one which we shall be looking at in a subsequent chapter.

Genetic variation is omnipresent at the level of proteins and this variability must have tangible effects at successively higher levels of morphology and behaviour. The extreme antibiologist has to defend his position with at least one of the following two claims, each of which appears unreasonable. He may first claim that the genes underlying the so-called higher human faculties are distinct from any of those for which a high degree of protein polymorphism has already been discovered by the methods of molecular population genetics. Secondly, should he concede any overlap here (and in our view it is likely to be a considerable one), then he may maintain that these polymorphisms will nevertheless not result in any tangible differences in individual behaviour. This last position can save at least the appearance of consistency only by being coupled with the view that the majority of alleles are selectively neutral. The final and most telling criticism that can be levelled against the antibiologist is his likely refusal to use the methods of quantitative genetics to test his position in so far as it concerns higher faculties' heritability in the second sense.

The overwhelming majority of quantitative genetic analyses of cognitive and personality traits attest to considerable heritability or genetic variation in such traits. (See, for example, Willerman (1979) for data and Royce and Mos (1979) for the theory of behaviour genetics.) This fact would not be altered even if it were found that the majority of all polymorphisms were selectively neutral. Our intuitive assumption that, for example, one's degree of neuroticism cannot be selectively neutral would not have to be altered, because it is not primarily the alleles investigated by the population geneticists that are responsible for the heritabilities of behavioural or other complex traits. They generally define a rare allele as one with a frequency above 5 per cent. Those with lower frequencies are not amenable to their analysis because that would involve unmanageable sample sizes. But it is exactly these very rare alleles that make for family resemblance and interfamily variation. With mutation frequencies in the range of 10^{-5} to 10^{-6} and a number of genes per individual in the same range, practically every individual carries some rare mutation peculiar to him alone. Furthermore, every individual inherits from his parents rare alleles peculiar to them, which make him resemble them more than any other members of his species. Many of these 'very rare' alleles are not selectively neutral – the impressive body of knowledge accumulated by behaviour geneticists attests to that. Those alleles which make members of one family resemble each other more than they resemble members of other families give rise to traits which in all likelihood are not neutral – such as personality traits, which quite obviously affect fitness decisively. To make an extreme point, people locked away in jails or mental hospitals leave on average fewer offspring than those not so isolated from members of the other sex.

The above argument is also unaffected by Feldman and Lewontin's arguments (1975) purporting to show that the analysis of variance model, the analytical basis of most quantitative genetic analyses, is inadequate to the task of demonstrating heritabilities of cognitive or personality traits. Our point could have been made without any heritability having ever been measured. All we need to show is that mutations have diverse pleiotropic effects on behavioural traits. Population and behaviour geneticists have clearly shown this to be the case for all animals so far investigated including human beings. We do, however, sympathise with Feldman and Lewontin when they question the sociopolitical and medical advantages to be gained from quantitative genetic analyses of cognitive traits. Such advantages have not been demonstrated yet, even if the demonstration of a heritable component in the aetiology of, say, schizophrenia could help one find a biochemical cure of that disease.

The danger that such knowledge will be used by politicians to suppress people of different colour or creed is very real. Not only did Hitler and his pedigree-fetishists use biological data (and bad ones at that) to further their own ends; but also more recently extreme rightists have invoked the names of Jensen, Eysenck, Dawkins, Maynard Smith and Wilson as supposedly corroborating racist and xenophobic attitudes. Jensen, Eysenck, Maynard Smith and Dawkins, upon being challenged publicly by a politically more active colleague, have forcefully dissociated themselves from the National Front and all other racist groups (see *Nature* 282 (1979) 348; ibid. 274 (1978) 738; ibid. 289 (1981) 742; ibid. 289 (1981) 528). It appears to be very 'human' to seek to rationalise emotional needs that conflict with moral codes; we are reminded of Hume's dictum that 'reason is the slave of the passions'.

The most elegant investigation to date of the genetic aetiology of human cognitive and personality differences is currently under way at the University of Minnesota in Minneapolis. Thomas Bouchard is seeking out and reuniting monozygous twins who have been separated from very early on in their lives. Many of them have been unaware of their twin's existence. Some anecdotal reports about this study have already appeared in the press: the similarity between such twins is breathtaking (Holden, 1980). Two British twins turned out each to have a cat named Tiger, and two children with almost identical double names; to like novels by the same author; to be wearing two bracelets on one wrist on arrival in Minneapolis, one bracelet and a watch on the other wrist, and seven rings. Objective science cannot say much about such astonishing coincidences, for it is difficult to assess probabilities of such circumstances. The twins have also undergone intensive physiological and psychological tests. In some the results were closer than is usually the case when the same person is tested twice. This was especially true for IQ. So far thirty-one twin pairs have been tested. When the study is over it will probably be the best answer to the nature-nurture question available. It already seems clear that nature's contribution to the 'whole person' may have to be emphasised more than is currently the case.

With twin studies what is frequently overlooked is that monozygotic twins are not, genetically, completely identical. So-called somatic mutations, occurring in the body cells early in embryogenesis, can profoundly differentiate the two individuals deriving from the divided fertilised egg. Also, if the two emerging cells do not share the fertilised egg's biochemical ingredients equally, they can easily yield individuals whose marked differences would certainly not be regarded as 'environmental' as the term is presently used in the human sciences.

Both these causes of differences between 'identical' twins are not very important as far as relative measures of heritability are concerned.

For both processes occur in dizygotic twins and therefore cancel out in the comparison. But when assessing degrees of similarity between twins it is worth bearing these possible causes of difference in mind.

Quite clearly some human characteristics are determined entirely by environment. What natural language a child speaks is in this sense very much an accident of birth. But what is at issue here, when considering the heritability of the language faculty, is not *which* language the individual comes to speak but *how well* he speaks it. This observation immediately raises three points. The first two are about language and are very controversial. The third concerns other human faculties as well, especially intelligence.

1 Are there objectively specifiable differences in individuals' mastery of language? Our whole educational system appears among other things to be concerned with improving such mastery and certifying the differences through examination. Yet it is widely alleged that formal education is blinkered and narrow, in neglecting how speech can be complemented with bodily gestures and attitudes, and in not taking account of taciturn individuals' resort to shared unspoken assumptions of subcultures to which they belong. The overexplicit verbaliser, it is held, is all too often – and mistakenly so – regarded as the star in a dull cast. Despite all this, there do indeed appear to be objective differences in such respects as number of vocabulary items, mean length of utterance, speed in disambiguation, discernment of logical connection and so on.

2 If these objective differences are conceded, then controversy flares once again when one asks whether the differences discerned are attributable to mainly genetic or mainly environmental causes. The differences may be correlated with, or be alleged to be correlated with, differences in age, sex, race or subculture. Those who hold the view that there is a genetic basis partially underlying these differences then lay themselves open to accusations of ageism, sexism, racism or chauvinism. This now leads to the final question, which we shall pose by straight away giving the answer.

3 It is impossible to infer from within group genetic variation causing phenotypic differences *within that group* any conclusions at all about the causes of between-group phenotypic differences. In order to be clear about these matters let us choose an example from outside

> the sociopolitical minefield. One and the same species of pine grows in the European plains and also in alpine areas. Differences in height among the plain dwellers (of the same age) can to some extent be attributed to genetic variation. Likewise differences in height among the alpine dwellers can to some extent be attributed to genetic variation. Overall the alpine dwellers are much shorter than the plain dwellers. But we cannot yet infer anything about the exact reason for this between-group difference in height, except to say that it *could* turn out to be wholly due to their different environments. This, of course, is the sanctuary sought by environmentalists in the IQ-race debate. They may be completely right; without yet falsifying the claim that within-group differences among individuals have a genetic basis. Although Jensen (1972) notoriously is at pains to present evidence and arguments for his view that there is a genetic basis for the difference between mean IQ's of different racial groups, these can never (as Jensen is, of course, aware) amount to anything more than plausibility arguments. This is because no matter how strenuously the nativist tries to show that members of different racial groups have grown up in the same environment (e.g. blacks and Puerto Ricans in New York slums) one can never exclude the possibility that some crucial environmental difference has gone undetected. It could turn on something as intangible as a confident sense of belonging to a well-knit social group. A similar situation prevails in the ecological debate over the so-called exclusion principle. This states that no niche can be occupied by more than one species. Purported counter-examples prompt ever more refined specifications of the niches in question, thereby saving the principle from refutation. The end point of such refinements may be such a narrow definition that no other species could occupy it – for the simple reason perhaps that no two distinct biomasses can occupy the same region of space-time. But with this extreme definition of niche the exclusion principle becomes vacuous.

The controversial claim about IQ and race would be similarly trivialised if the environmentalist resorted to overspecifying the environments in order to generate differences between them that could account for the differences in mean IQ. Such an extreme specification for example might be 'membership of the black minority within predominantly white America'. There is only one group in the environment thus specified, and *a fortiori* Jensen's claim cannot be tested by examining other groups within that same environment.

A final useful comparison in this connection is with von Mises's problem of collectives in actuarial science (1961). An individual's life expectancy on which, of course, his life insurance premiums depend, is determined by the mortality statistics for the collectives to which he belongs. The collectives are sets of people satisfying various conditions on age, sex, height, weight, daily habits, their own and their relatives' health, occupation, etc. In order to estimate the probability that a given individual will survive to a certain age, one needs mortality statistics covering as large a collective including him as possible. Yet one also requires other members of this collective to be as much like the individual in question as possible. The first requirement, therefore, is for a lax condition on membership of the collective; the second for a tight one. If one pushes the second too far, by specifying so many characteristics of the individual whose life is to be insured that only he satisfies them, then one reaches the absurd position of not being able to determine frequencies at all.

The tension between tight and lax specifications of environments and niches in psychometrics and ecology is therefore a tension between probable truths on the one hand and empirical generalisability or simplicity on the other. This is not without precedent in the philosophy of science; Sober (1975) has essayed on the pervasive tension between evidential support for, and simplicity of hypotheses in all branches of science.

Let us return to the problem of intergroup differences and their origins. Some population geneticists recently debated whether one could differentiate the human races on the basis of data on protein polymorphisms. Mitton (1977) applied a multivariate technique to such molecular data and showed that, taking a number of proteins, one could reliably assign individuals to their respective races. He was severely criticised for his mathematical methods by Lewontin (1978) among others. The data clearly show that the genetic variance in the genes investigated is much greater within each race than between the races. This does not, however, prove that there is therefore no significant inter-racial variance. As Neel (1981) has pointed out, a few genes can bring about considerable inter-racial variation. It appears, for example, that the difference in skin colour between Europeans and Africans is due to only about four genes. Other genes, perhaps few in number, could, of course, be responsible for differences in cognitive or personality traits between races. Certainly relatively minor inter-racial variance in polymorphism data concerning a limited number of genes does not exclude such a possibility. That there are many quantitative and few qualitative differences on the molecular level between races is an established fact. That multivariate techniques will therefore be able to assign individuals to their correct race is a logical consequence. The only interesting question is, how many of the quantitatively varying genes does one need to be able reliably to differentiate racial affiliation of individuals? This is an empirical question and Mitton has attempted to answer it.

To some people different races appear in all essentials identical and to others it appears as if worlds separate them. Both attitudes have nothing to do with objective science, and expectations flowing from them are wholly inadequate for the evaluation of empirical data. Humanitarian concern should not blind one to the fact that genetically races are not identical. What one might discuss is whether the discovery of genetic differences would be so dangerous politically that such research should be forbidden or at least not financed.

It seems that some people were just as surprised to find significant molecular differences between human races as others were when it was discovered that human beings and chimpanzees are, on the molecular level, almost identical (King and Wilson, 1975). In both cases the expectations of those surprised were entirely unreasonable.

Adorno (1950) is well known for his environmental explanation of human aggression as the legacy of authoritarian upbringing. Once more, of course, one could press the genetic question whether children from similarly authoritarian homes become similarly aggressive. Both matters are virtually impossible to measure and even comparing people in these respects is difficult. Whether aggression is easier to reduce by genetic or environmental manipulation is not obviously answered by any answer we may achieve to the first question. In a nutshell, environmental influences now acting on the child may themselves be due to its parents' genes; and attempts to change the former in the interests of the child's personal development may founder on the intransigence of the latter. Adorno is also guilty of a common fallacy: that of inferring causation from a simple correlation. Medical scientists frequently use demographic data to gain a better understanding of diseases. But one needs to establish further either one of two things:

(i) that experiments disclose a causal link;
(ii) that the genetic variable has been controlled. One cannot overlook the possibility that a given genetic constitution causes both

the habits and the diseases that are discovered to be correlated.

There is no link between amenability to therapy, on the one hand, and, on the other, the aetiology (environmental or genetic) of any behavioural disorder. Fighting fire with fire is an inappropriate rule of thumb in such complex matters.

It is jejune for the free will theorist, in seeking room for human free will in the causal web of doings, to set more store on either genetic or environmental causation than on the other. One might also mention here another common fallacy nowadays. This is the belief that emphasis on environmental influences promotes the prospect of utopian change – a change to be achieved by the collective exercise of individual and free wills. Skinner (1971), of course, is the environmentalist best known for his proposals for social engineering, the creation first and foremost of new environments or training programmes, to make people happier. He, however, can at least be exempted from any suspicion of such fallacy; for he is quite adamant that free will is a chimera with no role to play in the story of behavioural causation. The point is, however, that if the free will theorist seeks to usher free will through a back or side door as a late or uninvited guest, or tries to dress up one of those already present and pass it off as the force in question (compatibilism), he must contend with the fact that the guest list of causal agents contains *both* the environmental and the genetic. The question of free will aside, millenarian aspirations and ideals could just as easily be served by tackling the genetic system (selection or genetic engineering) as by modifying behaviour along Skinnerian lines. Extreme pan-millenarianism, as it might be called, might best be served by both environmental and genetic manipulation. One of the most important tasks for the former half of this programme would be to modify the prevailing *Zeitgeist* which sees genetic measures as the ultimate affront to that nexus of emotions and values attaching to the institutions of ownership and inheritance. Man has come to a conscious understanding of these institutions on the cultural side first, and the attendant emotions and values transfer with understandable vehemence to the genetic chattles that modern science is only now beginning to disclose. But once some form of Skinnerian training had softened up attitudes to genetic measures, perhaps the most effective steps down the road to Utopia would be measured in Ångstroms. Or so the pan-millenarian would have one believe...

But, of course, extremism is extremism in any guise. Granting the essential correctness of the point that both environmental and genetic causation is always involved, it is nevertheless too stark and simple-minded a response to think of the most effective utopian policy as a simple conflation of the two extreme measures – Skinnerian training and genetic engineering.

Our present, realistic joys and aspirations are what Utopia is surely made of – not some far-fetched end result of wholesale mutation and social manipulation. To this end, we would be far better off learning more about our *present* natures and creating realistic conditions for their happy fulfilment. We could best start by so changing our patterns of work and socialisation that we spent more time with our children; by so distributing society's material goods that no one was in abject need, or caught on a treadmill of consumerism. Such a list of modest proposals could be extended indefinitely; and ought surely to be the sanctuary of the sane reformer.

Let us now summarise our findings about heredity and environment. All that can be said about the nature-nurture problem is to be drawn from the following sources: Fisher's 'fundamental theorem of population genetics'; Waddington's account of canalisation in individual development (1953); and the data of modern population and behaviour genetics.

Any selectively important, continuously varying trait will
(i) lose much of its additive genetic variability in the course of a few generations (Fisher's fundamental theorem); and
(ii) be highly canalised ontogenetically, thereby always being adequately expressed over a wide range of environments.

Population genetics has revealed rather high mutation frequencies (comparable to gene numbers in higher organisms). Genetic variability is great in all species investigated so far. Behaviour genetics shows that almost all genetic mutations affect behavioural traits in manifold ways. All behavioural traits studied so far show genetic variability.

Thus one must expect genetic variation in any selectively important, continuously varying trait. Judging the selective importance of a trait can call for some discernment. For example, what language one speaks is of little or no selective importance. But what is important is how well one speaks whatever language one happens to learn.

Traits more closely bound up with an individual's ultimate fitness will be more closely 'screened' by selection for genetic deviations. Traits that cannot vary without adversely affecting fitness will be more highly canalised. Genetic variation will never disappear entirely. Nor can any system be so well buffered that genetic disturbances will not be expressed phenotypically. (Environmental disturbances and adaptive learning will be more thoroughly discussed in section 8.)

The reader might think these remarks too vague. We would claim in our defence that they still serve to clarify and advance the often confused discussion of the nature-nurture problem on the part of scientists and humanists alike.

One would hope eventually to be able to categorise traits with respect to their degree of canalisation, and genetic and phenotypic variability. At present, however, it is difficult to quantify them in a way that would permit comparisons. For example, human body weight can be said to be more variable than body temperature. Likewise male reproductive success can be said to be more variable in our species than female reproductive success. But only the latter comparison makes strict sense, in invoking the same standard of measurement. The former comparison makes some sort of intuitive sense, but really has no reasonable basis in any system of measurement. Comparisons such as these nevertheless abound in the literature.

Some traits vary very little – such as the number of fingers or number of eyes in vertebrates. But these traits do not vary continuously, and their genetic and environmental buffering is extremely solid. Variation here is only due to gross mutation in a few genes, or environmental disasters. All traits of major interest in psychology and sociology vary continuously, and for both environmental and genetic reasons. As far as the development of many cognitive and personality traits is concerned, the environment has been relatively stable over the last 10,000 to 100,000 years. There was ample time for canalisation to evolve (for those traits that were selectively important). Thus it seems likely that much of the variation in the latter would be genetic, and that they would not be much affected by normal environmental variations. Experimental evidence suggests that it is more difficult to evolve an efficient buffer against genetic modifications. (Remember the pleiotropic effects of practically all mutations investigated by behaviour geneticists.) This is probably because most mutations are too rare to allow the genepool to respond specifically to them. Environments are comparatively more 'predictable' for the genepool. Extreme sensitivity to environmental changes will frequently itself be due to genetic conditions. This is another reason why minority pathology reveals little about nature versus nurture in the majority of a population.

Unfortunately these speculations are extremely difficult to verify experimentally. Proper psychological tests are notoriously difficult to design, and experimentation itself introduces influences absent in real life.

7 HERITABILITY AND EVOLUTIONARY PAST OF A TRAIT

Natural selection tends to use up the additive variability in any species' genepool, as favoured alleles swell their proportional representation at their respective loci. The stronger the selective forces are, the faster this happens (Fisher, 1930).

Suppose selective forces focus on a locus with ten alleles. If a rare allele suddenly comes into favour as a result of changes in the world outside, it will rapidly displace all other alleles from the genepool. The genetic variability of the corresponding trait will be diminished to a level maintained by new mutations, while its phenotypic variability will be reduced to whatever level is maintained solely by environmental differences.

For this reason it is possible to make some inferences from the genetic structure of a trait to its evolutionary past. If a trait has negligible additive genetic variability (that is, heritability in the narrow sense) then, since there will always have been an ample supply of mutations at the participating loci, the trait must have been subjected to intense directive selection in the past. One would therefore be dealing with a trait that is or has recently been crucial for survival in the genetic sense.

As explained above, the additive component of genetic variance, considered as a fraction of total phenotypic variability, is called *heritability in the narrow sense*, whereas genetic variability considered in the same fashion is called *heritability in the broad sense*. The former yields a measure of the evolutionary potential of a trait. Investigations of numerous traits have confirmed these considerations. Characteristics subjected to strong selection pressure display low heritability (narrow) and are strongly determined by the environment. Examples are frequency of copulation in *Drosophila*, the speed of nerve impulses in mice and their ability to find an escape route from hazards like water, as well as those characteristics of farm animals that are direct indicators of fitness, such as the number of eggs laid by a hen, the fat content of cow's milk and the size of a sow's litter (reviewed by Schilcher, 1978).

Although the data are not at all satisfactory, human personality traits seem in general to have low heritabilities around 10-50 per cent (Willerman, 1979). Thus observed differences between individuals along Eysenck's well known psychological axes – introvert/extrovert, neuroticism/stability, and tough/tendermindedness – should be accounted for mainly by environmental differences (Eysenck, 1980). This would mean, according to the speculative hypotheses offered above, that in the past strong selection confined human personality types to within a relatively narrow genetic range. By contrast intelligence, as measured by IQ tests, has a relatively high genetic variability between 50 and 80 per cent (Bouchard and McGue, 1981). If most of this variability is additive, then again this would indicate that IQ has been of little selective importance in the past. Putting these observations together, we would conclude that past reproductive success has depended more on individuals' personality types than on their IQs. This implies *either* that the sort of rationality and cognitive ability emphasised by evolutionary epistemologists like Lorenz, if it has been of great selective importance, goes importantly beyond IQ, *or* that the story of the emergence of the 'ratiomorph apparatus' does not have the selective credentials claimed for it. Perhaps increased brain size along the phylogenetic lines leading to man has much more to do with the integration of mood, motivation, gregariousness, will to succeed and so on, than it has to do with the cerebral processes subtending IQ.

We must, however, enter a note of caution here about arriving at bold conclusions from such psychogenetic studies. The main problem is that we simply do not know to what extent the environment has remained constant. The variabilities involved are governed by the equation $V_P = V_A + V_D + V_E$. That is, phenotypic variability is the sum of three components: additive genetic variability, dominant genetic variability and variability due to the environment. A change of environment occurring between the era of selective significance and present-day measurements could have completely upset the relative sizes of V_A and V_E. A quantitative genetic measurement is strictly valid only for the particular conditions under which it was taken and for the particular population concerned. When we have no information about the history of the environment and the population, they do not help us to say what the dynamics of selection were.

Suppose that a selectively short time ago many new and interesting environments confronted man, hitherto evolving within a relatively stable one. This environmental proliferation could have brought to phenotypic expression a variety of genotypes whose differences had thus far been dormant. Total phenotypic variation would go up absolutely, and, as a proportion of this, so would heritability in the narrow sense. Indeed, this could happen without significant selection having had a chance to take place, because the environmental proliferation could take place within one or two generations. An inflated heritability factor would therefore mask the past directed selection that might have taken place for the trait in question. With rapid cultural diversification, and the growth of technology and education, this may well be what has happened in the case of intelligence. On the other hand the sub-environment most critical for the formation of personality – mainly the family unit in infancy and childhood – may, by contrast, have remained to all intents and purposes the same. These considerations would explain why today we observe generally lower measures of heritability for personality, but the contrastingly high measure for intelligence. This explanation in no way undermines the claim that intelligence, or whatever it is that IQ tests measure, has been strongly selected for in the past.

A further reflection on the simple equation above that is not without significance for social planning is due to Herrnstein (1973). He pointed out that making educational opportunities completely uniform and adequate to every individual's potential would, to be sure, decrease phenotypic variance in developed skills and capacities to the possible satisfaction of egalitarian policy-makers. But, very importantly, that proportion of this variance due to genetic causes would increase dramatically. And if, as is only to be expected, people continue to select as mates those of like minds and talents (so-called assortative mating) then society could stratify into classes that are much more sharply defined genetically than is the case at present. Moreover, members of society will accordingly be much less mobile between classes – a quite unpalatable result for the policy-makers in question. Huxley's alphas, betas, gammas, deltas and epsilons are an even starker possibility if everyone grows up in the same uniform housing estate and attends the same comprehensive school, all with classes of exactly the same size and teachers all of whom, of course, would by then be betas, with the same lower second class degrees obtainable from any one of the country's similar universities.

Lest the reader think that considerations from quantitative genetics have carried us away, we shall close with some careful reflections on the methodological difficulties facing that discipline. In order to determine heritability one must be dealing with a behavioural trait (since these are the ones we are considering) that can be measured quickly, for one has to measure enormous numbers of individuals to obtain reliable results. One human psychometric study (Loehlin and Nichols, 1976) used 800 pairs of twins and this was still not enough to give significant form to small differences in heritability. Moreover, one should set up the test conditions in such a way that they correspond as closely as possible to the natural conditions about which one wishes in a final analysis to be able to say something.

8 INSTINCT AND LEARNING

The discussion in this section is intimately connected with that of section 6 on heredity and environment. It makes certain points clearer and attempts to advance discussion of Lorenz's well known claims about instinctual components in human behaviour.

Every observer of animal behaviour will be struck by the way in which even very complex behavioural patterns consist of stereotyped parts that are executed mechanically without ever having to be learned. By contrast, everyone has once had painstakingly to learn skills such as driving a car or typing. The first researcher to characterise more closely what is meant by instinctual behaviour was Konrad Lorenz, famous as a founder of the European school of ethology, the systematic study of animals in their natural habitats. Every instinctual behaviour pattern consists, according to Lorenz, of a drive, an inborn releasing mechanism, very often a method of orientation and a consummatory act. The consummatory act is a sequence of movements which are performed *en bloc* in bringing the instinctual behavioural pattern to a close. For example, a frog may be generally on the lookout for a fly because he is hungry (the drive); when a fly is prominently available certain sensory imputs, upon filtration and collation in the frog's cognitive apparatus, somewhere strike a 'chord' in the frog's innate releasing mechanism. The frog now orientates himself for the consummatory act – the leap, the blow with the tongue and the swallowing motion. The consummatory act is almost always performed as a whole, even if the fly should move out of range at the very instant the frog launches into the sequence. The question of the extent to which the consummatory act is innate or learned has become a point of controversy between the Lorenz camp and a group of American comparative psychologists. A further contentious point, that of how much human behaviour depends on drives as postulated by some ethologists, will be discussed later.

The critical test to determine whether an action pattern is inborn is not its stereotyped nature but rather whether it is manifested before it could possibly have been learned. If, for example, a newborn chick pecks at small objects on the ground then this, for Lorenz (1965), is proof that the behaviour is inborn. Against Lorenz, Kuo (1970) maintains that since chicks perform rhythmic head movements in the egg, this behaviour after hatching cannot be inborn. Lorenz has countered by observing that some species of birds with very similar embryonic states do not peck after hatching but make gaping or sieving movements to obtain food. Kuo is just one of the many extreme environmentalists who offer hair-splitting explanations of all behaviour as learned or acquired. But of course not all nativists are exempt from such a charge either.

In discussing instinct we are, of course, concerned with the nature-nurture debate in its ontogenic aspect. We are no longer enquiring after phenotypic differences, but rather after the significance of genetic and environmental factors in the individual development of certain behaviour patterns. As already emphasised, no trait is wholly determined ontogenetically by only one of these two factors. But there are obviously quantitative differences in the respective degrees to which they determine a characteristic such as a hen's pecking at ears of corn. The difference in song development among fruit flies, cicadas and crickets at one extreme, and development in the repertoire and voice of a prima donna at the other, are so enormous that it is certainly meaningful to use different terms for the two extremes. The decisive difference, as Lorenz saw clearly, is that Drosophilas, cicadas and crickets brought up in isolation will sing their ancestoral roles perfectly, whereas a Kaspar Hauser prima donna would be somewhat disappointing in her debut as Madame Butterfly.

Between these extremes lie cases in which both factors are more or less equally important for the development of adult behavioural characteristics. The Lorenz school recognised early on that most forms of complex behaviour on the part of higher animals contain both innate and learned components inextricably interlocked. The shrike naturally stores prey insects on thorns; and even after growing up in isolation, will catch insects and try to impale them on any likely looking protrusion. He must, however, learn through experience that the best of these are sharp ones like thorns, and bits of barbed wire. Likewise, squirrels have to learn how to crack hazelnuts open as quickly and efficiently as possible, although they will automatically try to do this in an unskilled manner with their very first nut.

Chickens come to peck much more accurately in the first few days after hatching. This might be due not to a process of learning, but rather one of maturation. Hess (1956) had the ingenious idea of equipping chicks with displacing spectacles. The same concentration of their pecking pattern was to be observed even though they were now right off target. They could not have been learning to peck more accurately, for they did not have the benefit of any feedback. As this last example makes clear, it can be very difficult in particular cases to separate instinctual from learned components. (Matters are complicated further now by recent claims that attempts to repeat Hess's results have failed.) Nor do evolutionary considerations lead to any specific predictions. Evolution is an opportunistic business, choosing in each case a strategy that appears most favourable and is most easily achieved in the given genetic and environmental circumstances. The more complex the central nervous system is and the longer life-spans are, the more probable it is that the species will tread the path of learning. But this in no way implies that nematodes or paramecia cannot learn, nor does it imply that all human behaviour is learned.

Instincts importantly require drives in their definition. Another category of genetically determined behaviour is that of the reflexes. These have been divided by some theorists into the automatic and the conditioned reflexes, but this is a distinction that will not concern us here. Nor shall we be at pains to examine the complicated taxonomy of behaviours, superficially resembling learning, that include imprinting and habituation.

Eibl-Eibesfeldt (1973), a disciple of Lorenz, has studied many forms of human behaviour that are either reflexive or instinctual. It is well known that newborn babies will grasp at any object within reach, perform swimming movements in water, and go for the nipple within minutes of birth. Children born deaf and blind laugh and cry like normal children even though they could not have learned to do so – at least not in the normal way. It is more difficult to determine innate components of more complicated human behaviour given that Kaspar Hauser experiments cannot be carried out. In this case Eibl-Eibesfeld has recourse to a trait's universality as an indicator of innateness: thereby changing from the first to the second sense of heritable which we have been at pains to distinguish. The more universal a trait is across different cultures, the greater he, and many others, believe is the probability that it is inborn. This assumption is perhaps correct, but there is no conclusive reason to believe it. Firstly, gene frequencies can change much faster than many an ethologist supposes; certainly fast enough, for example, to give rise to genetically based differences in behaviour between Siberian and South American members of the mongoloid race, who parted company some 10,000 years ago. Moreover, there is no conclusive reason why learned forms of behaviour should not be homogeneous the world over as a result of convergent cultural evolution. We are, of course, assuming that cultural characteristics are subject to selection pressures. This will be defended in a subsequent section on cultural evolution. In order to compare the variance between cultures of a cultural trait with that of a biological one, one must, of course, know the selection conditions in their rough outlines. Unfortunately given present knowledge, one can give the rough outlines of these only for the simplest biological traits bound up directly with survival and reproduction. Even with so apparently unproblematic a matter as the number of descendants subtle difficulties are encountered upon close inspection. One and the same species can, depending on the conditions in which it is placed, become r-strategist or K-strategist. r-strategists place the highest priority on maximising their rate of reproduction. Such a strategy makes sense when entering a new niche in which there is hardly any competition from conspecifics or other species. K-strategists, by contrast, are favoured in conditions of evolutionary equilibrium under which it is best to ensure that descendants are as healthy and robust as possible in order to cope with primarily intra-specific competition. In this case, quality and not quantity of offspring is of the essence. Ethologists who make universality a mark of innateness in the ontogenic sense appear to be implicitly assuming that cultural characteristics are selectively neutral and consequently vary considerably. This is not altogether a correct view of the matter. As we shall see below, such traits are very probably also subject to selective forces. But even if this were not so, it would not follow without auxiliary assumptions that they would have to vary more strongly than biogenetic properties whose programmes are anchored in the DNA and not only in the nervous system. If the environment is heterogeneous, genetic variability could be greater than that cultural variability that has arisen through random events.

9 MICRO- AND MACRO-EVOLUTION, GRADUALISM AND PUNCTUALISM

A macro-evolutionary process is one such as the development from fishes to reptiles to mammals. A micro-evolutionary process is an alteration of gene frequencies such as is observed by population geneticists in the laboratory or by evolutionary biologists in the wild. Such alterations lead to relatively small morphological or behavioural changes within one and the same species. In the classical synthetic theory as formulated in its essentials by Dobzhansky (1970), Mayr (1963), Rensch (1960), Simpson (1949) and Huxley (1963), macro-evolutionary processes were very often regarded as extrapolations of micro-evolutionary ones. The latter are easily accessible experimentally. There are good mathematical models, for the case of two alleles at one locus, of the frequency changes that can ensue under various internal and external conditions (see Lewontin, 1974, for a review). Many experiments also have shown that under certain environmental conditions one genotype or another will come to predominate at the expense of others that differ from it in respect of one gene or gene complex (Dobzhansky, 1970). Some experimental work has gone so far as virtually to accomplish the process of speciation in the sense of reproductive isolation in the laboratory (e.g. Knight *et al.*, 1956). Different *Drosophila* populations, which once could be crossed easily with one another, have been subjected to intense selection in different directions, or simply been kept breeding apart, until they reached a point where hardly any genetic exchange took place between them when the opportunity arose. Thus they have, in effect, become distinct quasi-species. Dobzhansky's school of researchers have shown how *Drosophila* gene frequency changes in the wild occur in step with climatic changes. Furthermore, evolutionary processes in the wild have been directly observed. The most spectacular of these is perhaps industrial melanism, the example with which this book began (see Kettlewell, 1961). Moths' colouring changed across the generations to maintain their camouflage protection against birds while at rest on their favourite trees, whose colour was changing as a result of industrial pollution. What was happening, of course, was that uncamouflaged butterflies were picked off more easily by the birds, thereby increasing the representation of the genes for more appropriate – in this case darker – colouring.

There are, of course, no such observations to be had in the case of macro-evolutionary change. One has to resort instead to the more or less incomplete fossil record, a record which has always made paleontologists uneasy. Can factors investigated by geneticists explain the sorts of changes that the fossil record seems to speak of? In the last few years a group of American paleontologists (Gould, 1977; Eldredge and Gould, 1972; Stanley, 1979) have attracted public attention by once more questioning the theoretical edifice of the synthetic theory. Their arguments, however, are hardly new, and in fact raise doubts only about certain aspects of the overall theory. The fossil record reveals long periods of stasis in the history of many species and higher taxonomic groupings. These quiescent periods are punctuated by eruptive changes over relatively short spans. These evolutionarily brief moments of drastic and convulsive change have been dubbed 'punctuations'. Punctuation is maintained to be inconsistent with the Darwinian picture of gradual and continuous change of species. But Darwin himself had foreseen the possibility of varying rates of evolution. Moreover, the punctuating periods of, say, 50,000 years are, as far as population geneticists are concerned, ample enough to ensure considerable, even drastic, change if selective forces are strong enough.

The argument revolves in part around the old discussion of Goldschmidt's 'hopeful monster' (1940). He maintained that long ago that great evolutionary advances depended on the occasional production of macro-mutations with considerable morphological or physiological consequences. The new monster produced by such 'saltations' would happily turn out to be much better adapted than its normal fellows in some timely and crucial respect, and would accordingly set off an allelic landslide as its fitter descendants inherited the earth or whatever niche was in question. There is, of course, no question that such monsters could be produced. What is more difficult to judge, however, is how important a role they have played at the take-off points of many higher taxonomic entities. This is an empirical question and Gould and his colleagues have been gathering data in support of their punctualist view. It is quite unlikely, however, that macro-mutations resulting from single gene differences would be of positive evolutionary significance. As we saw earlier, such mutations are almost always lethal.

One punctualist (Williamson, 1981) has recently expressed bewilderment at having their ideas associated with Goldschmidt's. But in doing so he makes much of the breakdown of developmental homeostasis that is at the root of speciation events. Such breakdown is due either to drastic mutations (hopeful monsters) or drastic changes in the environment. So Goldschmidt's ideas clearly find an echo in punctualist doctrine – even if they disavow them. For if the punctualists are not saying that all larger morphological changes during evolution were induced by environmental changes, Goldschmidt's monsters rear their heads.

Consider the step from bilateral to radial symmetry in body design. It is as difficult to explain as a rapid response to a changed environment as it is by appeal to a single mutation. Of course, large morphological changes can result from small environmental differences. A famous case in point is the small nutritional difference that accounts for the difference between a queen and a worker bee. Here, however, we do not know the extent to which the rest of the genome mediates that differentiation. If the punctualist wishes to eschew mutational monsters then he should not let in preadaptional monsters through the back door. Environmental changes producing large developmental effects that turn out to be adaptive after such changes are no more likely than mutational monsters.

More plausible than hopeful monsters, however, is an old idea mainly due to Wright (1945) and Mayr (1963) and newly brought into the discussion by the punctualists: the early phase of a major new evolutionary direction is marked by great genetic upheaval, rather than a gradual genetic restructuring. Call a genetic outsider with rare alleles a *polygenic* monster, as opposed to Goldschmidt's monogenic ones. If it happened to be relatively isolated in a new niche where it happened also to be well adapted, and founded a new breeding colony that underwent intensive inbreeding initially, then the genetic balance that had been established in the original genepool would be disturbed. A new coadaptive genetic balance would quickly establish itself, and after a short time a new species could have been born that was genetically too incompatible with the mother species to allow further interbreeding. In geologically eventful periods, such as the end of an ice age, many small niches would have been opening up to provide opportunities for such genetic revolutions to take place. There is also some evidence that solar radioactive activity has been erratic in the past, and that a rich batch of mutations might thereby have arisen at times of peak solar activity. The scenario is graphically described by Valerius Geist (1978), who, curiously, is rarely cited in this context.

Those still active in the classical neo-Darwinian tradition imagine species originating as follows. Geophysical events separate a species in two, and the two halves breed for many generations independently. Both chance and environmental differences cause genetic differences to accrue on both sides, so that upon eventual reunion the hybrids produced between the two would find themselves at a selective disadvantage compared to non-hybrids. In both groups this would lead to selection for individuals disinclined to mate with members of the other group. Within a few generations of such a reunion two species could have arisen that are genetically isolated from each other. This is the picture of *allopatric* speciation. It appears that among the many species of Drosophila living in Hawaii, very rapid speciation has indeed been due to such accidental founder effects (Jones, 1981). On the other hand, there are also examples of the classical form of speciation (Bush, 1975).

The main point in favour of punctualism is, of course, the existence of gaps in the fossil record. This would be expected if the process of speciation were rapid and involved only a few individuals. Fossil remains from the period of transition would then be a very unlikely find. An archipelago like Hawaii offers ideal conditions for speciation according to the model of punctuated equilibrium. An organism like a fruit fly that can be borne on the wind across rivers, mountain ranges and between islands, is an ideal candidate for the kind of rapid change in which chance plays a greater part than selection. But, as Jones correctly observes, the very diverse vertebrate species that occupy whole continents lie somewhat beyond the scope of this explanation.

Whatever direction the debate may take over particular events of speciation and the formation of higher taxonomic categories, it is clear that all the models of such processes canvassed so far are fully compatible with the synthetic theory of evolution (Stebbins and Ayala, 1981).

The gradualism-punctualism debate, however, presents in an acute form the difficulty of apportioning causal credit to random genetic events (chance), on the one hand, and to law-governed selective necessity on the other. It appears to upset people imbued with a sense of purpose and dignity to suggest that man is a product of chance. This is particularly clear from discussions of Monod's *Chance and Necessity* (1971). He rather gives himself away by repeatedly emphasising the role of chance, as critics such as Eigen (1975) have pointed out.

Chance is part of, but by no means the whole story. A non-biological parallel might be illuminating here. Chance effects are congenially pressed into service by engineers seeking optimal solutions for complex problems. So-called bionic methods using electronic computers evolve solutions to engineering problems through the process of trial and error applied to conjectural solutions. Candidates are winnowed down by detected errors in the light of 'selection constraints'. The new conjectures subsequently generated tend to lie on a path determined by a short initial trajectory, for the time being foreclosing possibilities that could otherwise have arisen had earlier events been different. What transpires is that random trials with deterministic selection can produce optimal solutions to highly complex engineering problems that are not amenable to analytical methods in mathematics. What is more, some of the trials that the computer makes are like 'macro-mutation' conjectures, analogous to hopeful monsters in evolution.

Interesting in this connection too is Simon's notion of 'satisficing' in computational science (1955). He has shown that one can reduce lengths of search for solutions to computational problems by several orders of magnitude if one is prepared not to insist on an optimum solution but instead to make do with one within a pre-assigned tolerance thereof. (This process of near-optimisation he called 'satisficing'.) This has a corollary for evolutionary thinking. It is unlikely that species would achieve perfect adaptation even if environmental conditions were constant. Indeed, Sewall Wright's selective landscapes (1932) are an anticipation of this idea on the biological side. In his graphic metaphor, genetic constitutions may be thought of as points on a surface with hills and valleys. The hilltops are points of optimal adaptation with the genetic means at hand. The first rush up the hill is easy but the last few steps to the summit take real toil. Moreover, species set on a course but of a particular valley may find themselves stranded on a minor summit, as it were 'gazing' helplessly at much higher summits all around. To descend and then reascend a higher mountain would require too much dismantling of the genetic architecture. Unlike the grand old Duke of York, species are left enjoying whatever view satisfices. To add a further dynamic twist to this picture, one has to imagine environmental changes making the terrain bulge, sag, subside and shear. It seems to us that Sewall Wright's metaphor of the genetic landscape can be improved by imagining the landscape reflected in the vertical axis. Thus hilltops become sinks, and valleys become ridges. One can think now of a species' genetic constitution as a mass point on the surface rolling into sinks (states of greater adaptation) under the forces of gravity (selective pressure). According to Fisher's fundamental theorem (1930), the greater the additive genetic variability for fitness in the genepool, the less friction there will be to impede the motion of the ball to the bottom of the sink. It is now quite easy to appreciate how a ball with moderate friction could descend the side of a sink quite quickly, but slow up as friction increases (that is, as additive genetic variability is used up). This would make the last stretch to the bottom-most point, as the sink flattens out, more difficult to cover. Of course, when there is very little friction, the ball could charge down a long slope, pass through the minimum, and roll up an opposite short slope away from that adaptive sink altogether. Such 'overshoots' may very well occur with sexual selection, to be discussed below. To complete the analogy, macro-mutations would correspond to 'knocks' or impulsive forces on the ball, that could knock it from one general sink to another.

It appears unimportant to us whether chance resides only in mutations and recombinations as in the classical Darwinian picture, or was at work also through founder effects, and other punctuations.

As Edward Reed (1981) has observed, much of the excitement in the debate about the scientific status of Darwinian theory stems from a confusion of the notion of universal explanatory power and universal applicability. Natural selection is universally applicable in so far as no evolutionary process in the biosphere is imaginable in which it plays no role; but it does not have universal explanatory power which would allow one to predict evolutionary processes from the law of natural selection alone. As Reed explains, the situation is just the same for respectable laws such as Newton's universal law of gravity. Here other laws of physics also have to be taken into account when particular gravitational events are to be explained – most obviously one might need to know some other things about, say, aerodynamics.

'Panglossian adaptationism is the label Gould and Lewontin (1979), following Haldane, have attached to the position which claims universal explanatory power for natural selection. It is definitely a mistaken position but the law of natural selection does not therefore become invalid. Even extreme selectionists who expect an adaptive function behind each and every trait must admit that there might be some traits for which selection has played at best a minor role.

However, the pan-selectionist point of view is the only possible working hypothesis. This is simply because someone not searching for a function will never find one, even if one should exist. With all major traits that involve sizeable energy expenditure (in construction or execution) natural selection will always have been involved. The degree to which it has been is an empirical question that will be answered only by advances in developmental genetics.

10 ONTOGENY AND PHYLOGENY

No phylogenetic change can come about unless ontogenesis – that is, patterns of individual growth – alter. This is one of the most often repeated truisms of biology. Nevertheless very little is known about exactly what changes in ontogeny yielded new species, genera, etc.

The thread of our earlier discussion of micro- and macro-evolution leads us now to enquire further into the relationship between ontogenetic and phylogenetic change.

How far do morphological conditions prevailing within a species at any one time constrain and delimit the phylogenetic process? Epochs of phylogenetic stasis are perhaps due to morphology's stubborn resistance to the disturbances of genetic change. This hypothesis is not undermined by adducing the evidence of artificial breeding successes with domestic plants and animals, because, so one might argue, the modifications achieved are too small to amount to phylogenetic change.

No new full species has been created in the laboratory, apart from quite trivial cases where chromosomal doubling has achieved genetic separation between plant populations. Genetic isolation may be necessary but is not sufficient for the creation of a new species. The populations may be genetically separated but they are morphologically so alike that they do not pose anything remotely like the sort of phylogenetic change that separates a primate from an insectivore.

The punctualist further argues that the large changes needed to create new genera, orders or even phyla, cannot be achieved by genetic means alone. They are almost always a result of morphological alteration determined by physical conditions within the organism. Wilhelm His (discussed in Gould, 1977) showed a hundred years ago that many forms during ontogenesis could arise through differences in mechanical pressure in different parts of certain organs. For example, he used rubber tubes with elasticities like those of the embryonic brain tissue of vertebrates to model gross morphological development of the brain. Certain changes occurred only as a result of varying growth rates in different parts of the unfolding organ. It would be futile to seek the information regulating such shaping in the genome. It would only be found in the form of genes controlling growth rates. No plan would be found that in any respect resembled that made by an architect, enabling an artisan to construct a house.

Likewise a whole family of properties enter into simple allometric relationships. Growth rates of the embryo's unfolding tissues and organs are stably related to one another. Take, for example, an organism whose brain characteristically grows faster than its body. If its overall body size were to increase, then so would the relative size of the brain within the body. This would happen without any selective force for increased brain size and also without any specific genetic change being responsible for it. Thus one could, for example, explain why the human brain is so large without postulating specific mutations or selective forces. For reasons perhaps having nothing to do with selective advantages conferred by larger brains, the primates on our ancestral line could have developed ever larger bodies and enjoyed as a byproduct a relatively even larger brain suited to manifold uses. We are, of course, aware that brain size is not all there is to noûs; but the larger a brain is, the more likely it is that more complex neuronal circuitry can be forged.

General increase in body size can, but need not, have a genetic foundation. It is well known that in the last few decades average human body size has been on the increase. This is attributable not to selective processes but simply to improved nutrition and hygiene.

Some authors take these considerations as reason to revise the basic claims of evolutionary theory, or what they take to be this theory. Indeed, in population genetics the emphasis inevitably lies on small genetic changes, because these are frequent enough to be investigated experimentally. Whether great phylogenetic advances are due to a great many small mutations is nevertheless questionable. But that they should be is not a logical consequence of the discipline, nor is that discipline all there is to evolutionary biology. Classical evolutionists like Wright and Mayr have always emphasised the possibility of punctualism – even though the label is a new one. Modern critics like Lovtrup (1976), Gould, Eldredge, and Ho and Saunders (1979), are not really challenging the logical structure of evolutionary theory, but only redistributing certain emphases in application.

One of the key questions the punctualists are asking is the following: are there specific developmental genes which could be responsible for large phylogenetic steps? Could one find one or a few specific genes to be responsible for the change from, say, four to two wings in the evolution of insects? Dipterans (flies with two wings) are the more recent and advanced forms. They use a derivation of the second pair of wings as a gyrostabiliser, thereby achieving much better flight control. A mutation in *Drosophila*, a dipteran fly, restores four wings to the mutants. Here we have a mutation specifically involved with morphological development. This, of course, does not mean that the primary role of the protein coded for by this gene could not be located in the fundamental metabolic processes of the cell. But this gene clearly also plays some role in development. It could easily have been involved in the decisive phylogenetic processes that led from four-winged to two-winged flies. Other such mutations are rare. One of them makes a leg grow in place of an antenna in *Drosophila*. Others help determine the symmetry planes in wing construction. There are not very many others (Postlethwaite and Schneiderman, 1973). One may, however, safely assume that many such candidate mutations are lethal and therefore difficult to detect. Clearly this key question posed by punctualists will only be answered by hard and skilful work in developmental biology, at present not a very advanced discipline.

It is interesting that Stent (1980) claims that the kind of developmental genetics called for here is a futile hope. Stent is a molecular biologist turned neurobiologist. His criticism of developmental neurogenetics is a slightly clearer analogue of the punctualist's form of argument. He maintains that ontogenesis consists of programmatic phenomena, such as the production of amino acid chains. The base sequence of the DNA code is isomorphic to the amino acid sequence of the proteins. The very next step is the formation of a three-dimensional structure out of the amino acid chain structure, decisive for the function of a protein. But this step is no longer programmatic – for the DNA contains no precise description of this three-dimensional structure. The formation of this so-called tertiary structure is a typical epigenetic phenomenon, with genetic instructions finding their expression under the physicochemical conditions in the cell. But it is a fact that in conditions normally prevailing in a cell, a primary structure (amino acid sequence) always results in the same tertiary structure (3D protein). It is a further fact that specific base sequences have not arisen in order to code for the primary structures concerned. They have arisen rather in order to create tertiary protein structures fit for certain functions eventually to be evaluated by natural selection. One can manipulate the tertiary structure of a protein by changes in the base sequence, as well as by changes of pH value or of the temperature of the solution in which the protein is synthesised. In more complex morpho-genetic processes more genes are involved, and they depend on very many, frequently unknown, environmental parameters. But as we have just seen in the model case of tertiary structure of proteins, this does not render morphological processes 'far' from the DNA level inaccessible to changes inflicted by mutation. It is no exaggeration to say that each and every conceivable biological structure or function can be modified, though not necessarily arbitrarily, by changes at the DNA level alone. The great problem here is that every gene will have manifold effects (pleiotropy), and tracing and classifying these will be fearsomely difficult in a field like neurobiology. That, however, does not justify a defeatist point of view such as Stent's.

Webster and Goodwin (1982) complain that the Darwinian paradigm does not explain how functional and structural complexity evolved. Weismann's distinction between germ line and soma is allegedly a mystical partitioning: and what is missing are laws of form to explain how one structure is transformed into another without any interference from the 'central directing agency' of the germ line.

Now of course the Darwinian law of natural selection falls short of a full explanation of developmental processes. Something like the harmonic functions that Webster and Goodwin propose might help to tell which processes are possible, and which impossible. But such laws of form merely embellish the paradigm. We would not have to modify our view of the role of DNA and proteins in organisms. DNA would still be the prime matter of evolutionary change. There is not much difference between a rare mutation that allows further adaptation, and a widespread pre-adaptation of individuals that allows them to alter forms or functions when new circumstances demand that they should. In the first case DNA triggers morphological or physiological modifications in a constant environment. In the second case DNA provides the potential for a modification triggered by the environment. In either case the modifications depend on the information in the DNA. Darwin cannot be blamed for shortcomings in our understanding of individual development. Otherwise one might as well complain that the law of natural selection does not fully explain the physiology of developed organisms. The Darwinian law is a covering law to be supplemented by others describing processes within cells and within developing organisms. One does not rely on Newton's laws of motion alone to predict the pattern of vibration of a plucked string.

No discussion of the relation between ontogeny and phylogeny would be complete without mentioning the so-called biogenetic law due to Haeckel, Darwin's most important German populariser. This law says that ontogeny recapitulates phylogeny. According to Haeckel, the phylogenetic development of higher organisms comes about through accretion of further, terminal phases of ontogenesis while earlier phases remain intact. This implies a constant increase in the speed of development through these phases, given that more and more of them have to be embraced within a relatively constant embryonic span. Haeckel regarded the gills in human embryos as traces of our piscine ancestry, and provided hundreds of other examples besides. His law has since been discredited by many exceptions. Modern genetics reveals that developmental tendencies can slow down as well as accelerate during phylogeny. New traits can accrue at any ontogenetic stage and not all old traits have to be maintained in the way he envisages. In the case of our species in particular, a most important factor has been the decrease in the speed of ontogeny, making us the slowest developers among the primates. We need the longest gestation period and yet are born prematurely. We also have the longest pre-adolescence, maintaining youthful inquisitiveness and learning ability well into maturity. Gould (1977) has recently written an excellent book about the historical background, social effects and biological validity of Haeckel's law. Ontogeny can clearly disclose something about phylogeny. Perhaps Haeckel's maxim should not be treated as a law to be demolished by exceptions, but rather as an interesting frame through which to see biological data when seeking the best explanation of them.

11 THE SPEED OF EVOLUTION

Non-biologists are often prone to the impression that evolution is an inordinately slow process that is measurable only in millennia, at least as far as man is concerned. This is simply wrong. To be sure, evolution can proceed slowly or even come to a standstill, as it has with certain molluscs that have remained unchanged for 500 million years in the ocean depths. But in other cases it can take place very fast indeed. Horses have, over the last 70 million years, cantered through eight families, ranging from the four-toed Eohippus, which is no larger than a rabbit, to the animal we are familiar with today. Industrial melanism is another case in point, cited earlier in section 9.

Since the fifties, behavioural geneticists have published a flood of articles testifying to successes in artificial selection for various behavioural traits. Only very few traits, such as increase in mating speed in male *Drosophila*, have resisted. Presumably these traits have been subjected to constant natural selection in one direction (such as high speed) and all their additive variance has been used up. Rats, mice, rabbits and *Drosophila* have all had their motor, sexual and learning behaviour altered in all conceivable respects by artifical selection under the most varying conditions. (See Fuller and Thompson (1978), the standard textbook of behaviour genetics.) Depending on the intensity of selection, the available genetic variability and the trait chosen, the behavioural feature concerned can be drastically modified within three to six generations. With *Drosophila*, the effect has been achieved of dividing a population of flies that once mated indiscriminately into two sub-groups whose members will practically never mate with one another (Knight *et al.*, 1956). This is, of course, a case of artificial selection, but the same holds for natural selection in the wild.

In order to emphasise the dramatic effects of selection we shall now describe in some detail an elegant experiment performed by Dobzhansky and Spassky (1976a, 1976b). They wished to study the effect of social mobility. They subjected two groups of fruit flies to selection in opposite directions, calling the one group 'Aristos' and the other 'Plebs'. The selected traits were positive and negative geotaxis – simply, a tendency to go up or down. The first Aristo generation contained twenty-five males and twenty-five females. The founder Plebs numbered 500. In each generation the ten best Plebs (those with the strongest urge to walk downwards) were promoted to the Aristo ranks. These Plebs were the 'best' in the sense of the direction in which *they* were being selected – so in fact they were the ten Plebs *least* like the Aristos they were to join (for the Aristos were selected to move upward). The Aristos in turn sent their ten worst members to join the Plebs – these ten now being those among the Aristos *most* like the Plebs they were to join. Now *Drosophila* populations can multiply several hundred-fold with each generation. A representative sample of 600 was chosen from among the first-generation Aristos, to be tested for the trait concerned. The forty best were then used to found the next generation, along with the ten new recruits from the Plebs. From the first generation Plebs, 600 were likewise sampled with the 490 'best' (i.e. most unlike the Aristos) going forward with the ten Aristos 'twits' to make 500 altogether for the next generation.

A non-geneticist would now, perhaps, expect the two populations simply to change in their respective 'directions' of selection, possibly with slightly different rates. The small group of Aristos would change quickly as a result of intense selection, while the large group of Plebs would do so slowly because of weak selection. But what happened instead was the following. The elite population reached their goal of geotaxis very fast, essentially undisturbed by the ten 'sensations' on each breeding session's 'honours list'. But the behaviour of the large Pleb population was not modified in the expected direction. They changed in the same direction as the Aristos, and did so because of the rejects from the Aristo population that they recruited in every generation. Now, of course, flies are not people. This example, despite the suggestive labels used, is intended only to illustrate the power of selection, not to shock those who wish to believe that all men are much the same as far as their genetic endowment is concerned.

Geotactic behaviour in *Drosophila* has a heritability of only 3 per cent. Despite this, one can change it so drastically by selection in different directions that after two to four generations the two groups thus differentiated no longer overlap in test results. Only about twenty generations are needed to ensure that the trait is manifested as fully as possible, or not at all. It is worth remembering that, depending on the author, the heritability of human IQ is put at between 50 and 80 per cent. Artificial selection would no doubt achieve startling results.

If geotactic behaviour in *Drosophila* were 100 per cent heritable (in the second, narrow sense we have so often emphasised) – that is, if all differences in this respect between individuals were attributable to their genes – then one ought naturally to expect a quite different result from the experiment by Dobzhansky that we have just described. For in that case the Aristo 'twits' would be genuine genetic failures, and would contribute no 'worthwhile' geotactic genes to the Plebs. It is only when heritability is low that phenotypic value is no good guide to genotypic value.

12 THE DIRECTION OF EVOLUTION

Many have regarded evolution as proceeding from 'lower' to 'higher' organisms, as possessing an overall direction through proliferation: possibly even directed towards an end state of perfect and harmonious co-adaptation, distinguished by a particular species that would be the culmination of all that had gone before. Teilhard de Chardin (1959) happily has it so, in a paean to the perfectibility of the cosmos. The result is little more than mystical obfuscation – such bad science that it is not just science that is no good, but is not science at all. We would not wish to deprive the reader of his first encounter, should he not yet have made it, with Medawar's devastating review (1961) of Teilhard's book *The Phenomenon of Man*; but cannot resist the temptation to quote at least this much:

> How have people come to be taken in by *The Phenomenon of Man*? We must not underestimate the size of the market for works of this kind, for philosophy-fiction. Just as compulsory primary education created a market catered for by cheap dailies and weeklies, so the spread of secondary and latterly of tertiary education has created a large population of people, often with well developed literary and scholarly tastes, who have been educated far beyond their capacity to undertake analytical thought. It is through their eyes that we must attempt to see the attractions of Teilhard, which I shall jot down in the order in which they come to mind . . . 1. *The Phenomenon of Man* is anti-scientific in temper . . . 2. It is written in an all but totally unintelligible style, and this is construed as prima facie evidence of profundity . . . 3. It declares that Man is in a sorry state, the victim of a 'fundamental anguish of being', 'a malady of space-time', a sickness of 'cosmic gravity'. The Predicament of Man is all the rage now that people have sufficient leisure and are sufficiently well fed to contemplate it. . . . I have read and studied *The Phenomenon of Man* with real distress, even with despair.

Can one say that man is in evolutionary terms higher than a mouse, a roach or an orang-utan? Biology does not readily yield an answer. If 'higher' is to mean better adapted, one faces the problem of how to judge different organisms' adaptedness to their respective natural environments. Is an earthworm better or worse adapted to life underground than a shark is to life underwater? Is a mouse better adapted to life behind the wainscot than a roach? If 'higher' is to mean more prolific in breeding, then the orang-utan would certainly be the lowest of the three species first mentioned. If 'higher' is to mean more independent of its physical environment then we are certainly higher than the orang-utan which is in turn presumably higher than the roach. But here it is quite obscure exactly what features of the environment have to be considered when judging independence. Finally, if 'higher' is taken to mean having better survival chances then the question can be consigned unanswered to the waste basket.

Despite these complications, *inorganic* evolution clearly proceeds in a direction of increased complexity. Hydrogen, the simplest atom, was produced first, with elements of greater atomic number appearing later. Thereafter came molecules, the large ones placing us at the prebiotic threshold. Parallel to this formation of the building blocks of matter there was macroscopic evolution too, as galaxies were formed containing solar systems that in turn contained planets. *Organic* evolution seems also, despite the difficulties already mentioned, to have delivered greater complexity; it has certainly produced more DNA, and the biomass has increased. The question is now whether this trend must continue or whether there could be a sudden about-turn.

Physicists give the clear answer that complexity can only increase overall. (There may of course be local retrogradations, as in the oft-cited case of certain parasites that 'lost structure' in slinking out of the limelight back to a simple life.) Bronowski (1970) has offered a good qualitative account of all this with his notion of stratified stability. Forces see to it that things dispose themselves in a hierarchy of stable assemblies. For example, when two hydrogen nuclei collide they can form a helium nucleus. But then atomic forces ensure that three, rather than two, helium nuclei are needed to yield the next stable level, that of a carbon nucleus. The same development can be recognised in the organic order. The complex molecules forming part of a hypercycle, defined below, eventually arrange themselves in the stable form of a simple cell; and hierarchical composition proceeds through yet more intermediate stages – from multicellular organisms to nation states. Just as accidental collisions of simple atoms take one over the threshold to more complex atoms so also accidental mutations take one over evolutionary ridges to stable dells represented by new and more complex organisms. Evolution can thus be seen as a climb up a mountainside with resting points in sheltered spots; a ratchet-like ridge with stopping places separated by slippery slopes.

It was the physicist Erwin Schrödinger (1944) who first made the point that organisms 'feed on negentropy' in order to prevent thermodynamic decay. Indeed, not only do they prevent their own decay by using energy coming in ultimately from the sun, but their genetic information is actually increased from time to time in the stepwise 'ratchet-like' manner mentioned above. Accidental shifts to higher levels of stable complexity occur with a long-term probability approaching certainty. Thus Bronowski: 'What evolution does is to give the arrow of time a barb which stops it from running backward; and once it has this barb, the chance play of errors will take it forward of itself.' Bronowski's essential advance over Schrödinger's resolution of an alleged incompatibility of evolution and thermodynamical decay is to have pointed out the possibility of preferred or stable kinds of structure which provide new cluster points for the random variation of energetic systems in the long run.

In physics the theory of nuclear forces has managed to account for why it is that there is an end to the production of ever-heavier stable atoms. Likewise, it seems to us, it should be regarded as a presssing theoretical task for biology to characterise the physical processes underlying organismic stability, to help us form some clear idea of just how far stable complexity can go. Might we already have reached a biological stage analogous to that of transuranium elements with a relatively short half-life? Eigen and Winkler-Oswatitsch (1981c) emphasise an important point in valuably extending Bronowski's idea. This is that a structure's information content can be stable only if maintained below a certain threshold. This threshold is inversely proportional to the average rate of error (that is, mutation) in the replication of individual structures or organisms. Note that Bronowski's idea straddles both the inorganic realm (where there is no replication) and the organic realm (where there is).

So long as environmental conditions remain constant, biological development cannot be reversed; although the rate of change could be slowed down in an asymptotic approach to a steady state. Spiegelman (*et al.* 1968) has already demonstrated this with DNA molecules as 'organisms'. Still assuming a constant environment, higher strata of stable complexity would have to be regarded as lying above this asymptote.

But, of course, no environment ever remains constant. Even if identical inhabitants merely increase their number, this will constitute a change in the 'environment' of each of these. But more importantly, accumulation of metabolic end-products will ensure significant change in those physico-chemical conditions of life on which evolutionary enterprise funded by mutation can capitalise. An example of decisive environmental change as a result of metabolic processes is how (according to some theorists) the earth's atmosphere changed from a reducing to an oxygenising one as a result of the first photosynthesising organisms. This change literally breathed further life into our planet. Kiefer (1979) has advanced an enlightening explanation of why organic evolution will always lead to greater complexity in the manner captured qualitatively by Bronowski. In thermodynamic terms, the fittest organisms are those that most efficiently exploit their resources. Efficiency increases with the structuration of the organism. The more compartmentalised the organism is, the easier it is for it to bring chemical reactants together at the right place and time with minimal heat loss, and the faster can those reactions proceed whose union of reactants is not left to chance.

But these considerations do not enable us to make precise predictions about the subsequent course of evolution. As happened in the Permian age, the earth might lose some of its higher stable levels of living forms through wholesale extinction of species. Thereafter progress would have to pick up from the highest level remaining until it once more attained a plateau. This plateau need not, of course, be the same as earlier ones because chance always enters decisively in blazing the exact trail. The presence of the right mutations at the right time is, of course, entirely a matter of chance. In this respect, evolution is like all historical processes in not being fully determined by law. Those laws that do hold are statistical in nature.

13 NEUTRALISM VERSUS PAN-SELECTIONISM

Will any environment so shape each species within it that a maximally adapted type will eventually emerge, deviations from which will appear as mutative or environmentally caused pathologies? Until the middle of this century, many biologists had it so, and consequently believed genetic variation to be very low. The true extent of genetic variability has only become known in the last few decades. The method of electrophoresis has been invented, enabling one to differentiate proteins by shape and charge. Experiments can now separate proteins of different electrical charges. Since the sequence of amino acids in a protein matches (is colinear with) that of the base sequence in DNA, electrophoresis gives indirect access to DNA. This method has revealed that about one in three human genes is polymorphic – that is, has at least two alleles occurring with a frequency of more than 5 per cent. On average each of us is heterozygous at one in ten loci. Other species display a similar degree of polymorphism. The variability revealed by electrophoresis is a lower estimate since it by no means discriminates all features of protein molecules that may be of selective significance. Variability could actually be two to three times greater. Indeed, recent work making use of improved electrophoretical methods has revealed considerably more polymorphism in enzymes. Enzymes that appeared to be homogeneous on earlier analysis now turn out to have ten or more variants. Enzymes with twenty variants are not exceptional. One must, however, add that this does not appear to hold for structural proteins, which are by far the most frequent ones and whose function to a large extent depends on their three-dimensional structure. In their case, variants in the amino acid sequence are the exception rather than the rule (Jones, 1980). We must repeat here that these investigations say nothing about very rare alleles. These exist in great variety and can be responsible for much phenotypic variation; but they cannot be detected in manageable sample sizes.

Experts are divided over how significant this variability is. The more theoretically inclined calculate that the polymorphisms have no selective significance – that they are neutral (Kimura and Ohta, 1971). The more experimentally minded prefer to seek mechanisms that might explain this unexpectedly high variability, and to design experiments that would show whether it is selectively significant (Clarke, 1970).

Lewontin (1974), who was one of the first to discover electrophoretic variability of proteins, has called for a new discipline to consider certain factors overlooked in the traditional one. One of these is gene linkage, associations that resist recombination in a way that permits groups of linked genes to become units of selection. Another, just as important, is pleiotropy. The manifold phenotypic effects of single genes certainly does call for closer examination. But a great part of the store of genetic variability would have to be neutral before depriving behavioural variation of its genetic basis. The selective effectiveness of numerous protein polymorphisms has already been established experimentally (Ayala, 1974). Moreover, as we know from animal breeding and behaviour genetics, there is hardly any trait that cannot be altered by artificial selection and this means that genetic variability must be available. We also know that all interesting traits in this connection are polygenic, so it is highly probable that at least some of the genes determining any trait are polymorphic.

If most polymorphisms are maintained by natural selection after arising by chance (mutation), then one ought to be able to indicate what mechanisms ensure this variation. Classical evolutionary biology sees selection as working constantly to reduce variability, with existing mutants being attributed to recurring mutations and indeed occurring with very low frequency. The mutation rate of single genes is of the order of 1 in every 10^5 to 10^6 events of replication. The most important way in which selection might maintain variation is, according to pan-selectionists, superior heterozygote fitness. It is easy to show in this case that both homozygotes at a locus with two alleles will always be reproduced, despite being at a selective disadvantage. A further mechanism is frequency-dependent selection of alleles. That is, as an allele increases in frequency its selective advantage decreases. For example, females might prefer rare males as mates. An analogous effect results when predators prefer common to more deviant prey. To continue the list, variability can be maintained by selection if members of one species inhabit different parts of an heterogeneous environment in which they are respectively better adapted than their fellows elsewhere, but can migrate from one part to another. Finally, cyclic and disruptive selection can also underwrite part of the insurance policy of variability.

Some theorists claim that the amount of variability existing in natural populations is clear proof that most of these polymorphisms are neutral in their effects on their carriers. They claim that no population could afford such high numbers of detrimental alleles in their genepool: because for an allele to be removed, an individual carrying it has to be removed from the breeding population. They show that under certain assumptions so many individuals would have to be removed that the species would become extinct. Some of these assumptions are, however, quite arbitrary and, as Lewontin (1974) showed, different numbers in the premises make the genetic load acceptable (as this removal of the less fit from the breeding population is called). The higher the load, the higher the frequency of advantageous mutations. Somewhere each species must find a compromise. Most species have similar mutation frequencies.

14 SEXUAL SELECTION

Darwin realised that a selective force could diminish males' survival chances yet increase their reproductive success: especially so if males, in order to increase their mating chances, develop, say, long and resplendent tails – like peacocks or birds of paradise – despite the apparent handicap to survival that these adornments entail. R. A. Fisher (1930) later made qualitatively clear, as did O'Donald (1962) quantitatively, that if females start choosing as mating partners males with longer tails and if, in the beginning, there were a correlation between tail length and fitness, then a 'runaway' effect could result. Males with longer tails would be more successful as courters, and females choosing them would in turn produce sons similarly endowed and hence also successful. Their daughters would, of course, guarantee that success by having their mother's preference. This process could reach a stage where tails became real handicaps to males outside the domain of courtship and mating – thereby reducing their overall fitness.

It seems to us, however, that this would lead to an unstable situation, which would be vulnerable to a female mutation preferring shorter tails, or choosing a different mark of male suitability. The male offspring of such a female would thus have higher overall fitness.

A typically maverick suggestion of Zahavi (1975) is that the handicap is a 'principle' that ensures fidelity of male fitness-signalling. The male is 'saying': 'Look, here I am, lumbered with this great tail – I must therefore be very fit in order still to be around.' The female demands from the males the elaboration of such adornments as imply a real survival handicap because otherwise she would be susceptible to 'cheating' – that is, to deceitful signals of fitness. If the long tail were no handicap, then basically unfit males could be as successful as fit ones simply by the *faute de mieux* of sporting a long tail. The female would in such a case have to look for other more reliable characteristics in appraising the fitness of potential sexual partners.

But if put to close mathematical test, Zahavi's handicap principle does not stand up: basically because a genetic handicap of any kind to pre-breeding survival chances will always be counter-selected (Maynard Smith, 1976b, Davis and O'Donald, 1976). It is worth mentioning here that the fact that females do choose from among courting males is well established for many species (Campbell, 1972). Whether the apparent handicap really is a genetic handicap in the above sense, however, is not easily established. It might well be that what appears to us as a real threat to survival is no such thing after all.

It is most likely that each male develops a tail of such length and splendour as will just fail to impose a handicap (by, for example, attracting predators' attention, reducing his mobility, etc.). Nevertheless, the quality of the sexually selected characteristic would be clearly correlated with the overall fitness of its carrier. There are good reasons to assume the latter – birds with greater metabolic efficiency, greater discernment in diet, etc., can 'afford' certain 'luxuries' morphologically. Therefore females are able to make reliable choices of fit males on the basis of their adornment. Furthermore, this adornment will not impose a handicap unless some male 'cheats' and tries to display more splendour than it can afford – in which case the abovementioned counter-selection would restore its balance. Sexual fitness is thus an entirely normal component of the fitness of any animal.

It should not be thought that all sexually dimorphic characters are the result of sexual selection. Other reasons could be, for example, purely intra-sexual competition. Males compete fiercely for females and thus become heavier, develop weapons for fighting, etc.

That females in our exposition of sexual selection have been the choosers, and the males the ones who compete, is due to no expository whim on our part. This is how things usually are with vertebrates. Trivers (1972) has given the reasons for this, developing ideas put forward by Haldane, Fisher and Bateman. In his theory of parental investment he shows how the sex that invests more in the offspring from gamete onwards is more likely to be discriminating in choice of partner. The sex that produces large numbers of small gametes and takes no part in the rearing of young must have as its aim to fertilise as many members of the opposite sex as possible. This is so because it has relatively less to lose by begetting less fit offspring.

In all animals, the females by definition produce larger gametes than the males and in many species go on to care much more intimately for the developing young. The result is different strategies for the two sexes. A female potentially produces fewer young than a male and should accordingly try to choose males that are as fit as possible. Thus she drags out courtships in order to find out all she can about the males' fitness. The male, on the other hand, should attempt to beget as many young with as many females as possible since each mating costs him but little of his future reproductive potential.

This generalisation holds in the extreme form only when the male reproductive role is limited to insemination and does not extend to fathering in the full sense of that word. But even in man the general tendency is there. The theory was beautifully corroborated in the case of fish with greater male postnatal investment in offspring. Here the female is the fickle one, and the male the coy and choosy one.

Darwin saw clearly that sexual selective effects would be all the more marked the greater the variance in reproductive fitness among males. Thus in polygamous species, the evidence for sexual selection is more striking than in monogamous ones.

Let us repeat here that sexual selection is conceptually on a par with group and individual selection. It results from the particular environment that sexually reproducing organisms find themselves in; an environment in which ability to compete for mates is an important part of an individual's selective value. One cannot see it, as many have done since Darwin's days, as in any way countering allele selection (or, which is the same, natural selection). Just as with group selection, the new situation is startling because it results in selective pressures that on a naive first look seem to militate against the interests of the individual. With group selection the individual can forego reproductive chances; and in the case of sexual selection it may put its survival at risk for the sake of its reproductive interests.

The fitness of an individual is a composite sum of various selective components. Good survivors and good reproducers can both have low fitnesses – the former if they do not reproduce, and the latter if they do not survive. So, clearly, it should cause no surprise if sexual behaviour sometimes seems to put individual survival at risk. It was surprising to early evolutionists because they put too much emphasis on survival, as their catch-phrase 'survival of the fittest' indicates.

15 LAMARCKISM

If one acquires callouses in one's work, can one expect that mark of honest toil to be present from the start in one's offspring? One of the first evolutionists thought that characteristics acquired in one's lifetime could be passed on to one's children not just by teaching or example but by a process that transforms acquired information into stable genetic versions thereof. It is not only individually learned information that can be transmitted in this way. It is also the physical result of habitual exertions such as the proverbial longer neck of the proto-giraffe striving for the greenest titbits on high. The mechanism of natural selection, as we have seen, is powered by a genetic engine that coughs and splutters, without regard for where one ought to go, and is steered by selective forces. But the mechanism of inheritance of acquired characteristics proposed by Lamarck (1809, trans. 1914) would make evolution a much more streamlined affair. To continue the metaphor, every favourable turn of the wheels would open the throttle further. Darwin, in his unfortunate ignorance of Mendel's results, toyed with the idea of a Lamarckian mechanism of evolution. But, except for some occasional resurgence of interest in Lamarckism, Darwin's mature theory of evolution by random variation and selective retention holds sway. There are two main reasons for this. First, experiments specially designed to test the two hypotheses have clearly favoured

Darwin's. Second, the synthetic theory has provided a good micro-model of the underlying physicochemical constituents and processes that make up the Darwinian mechanism. Moreover, in the micro-model there hardly appears room for Lamarckian 'reverse translation'. Such a micro-account is conspicuously absent in the Lamarckian case, although we shall speculate about some possibilities below.

First we shall explain the experiment designed by the Nobel laureate Lederberg and his wife (1952) that speaks positively for the Darwinian mechanism. What it shows is that if a population of bacteria develops a genetic resistance to a certain virus, this resistance must have been present in at least one member of the strain before exposure to the virus. This is not explicable by appeal to ancestors' developing the resistance by their own individual Lamarckian strivings. If one cultures bacteria on a plate in a nutritive medium containing the virus (or an antibiotic), one finds that some of the individuals are resistant to it and can multiply in its presence. Having stored the forebears of such individuals, one can then see whether these already had the resistance, as Darwinism maintains they would; or whether they did not, but would develop the resistance if treated with the virus. In fact, the former is the case. The forebears of resistant individuals are also resistant, even though they have never had contact with the antibiotic. The possibility must, of course, already have been excluded that this heritable resistance had been produced in a Lamarckian fashion by earlier contact in previous generations with the antibiotic in question. Such experiments have been repeated in various forms and they have always yielded the same result.

A decade ago Arthur Koestler (1971) tried to rehabilitate Lamarckism with his interpretation of experiments in so-called genetic assimilation by Kammerer and Waddington. Some of Kammerer's experiments with the midwife toad are now known to have been faked and we shall not describe them here. Waddington's (1953) were as follows: *Drosophila* larvae subjected to heat shock frequently develop an abnormality called 'loss of crossvein'. If one selects in each generation the most severely afflicted individuals for further breeding and repeats the process for several generations, flies are produced displaying the abnormality without having been subjected to the treatment as larvae. It is all too easy to see this as a case of Lamarckian inheritance of an acquired characteristic, albeit an abnormal one. But the Darwinian explanation is quite straightforward. In choosing the worst afflicted individuals in each generation, one is *ipso facto* choosing those with the lowest tolerance to heat shock. Artificial selection is thus lowering this threshold to the point where the abnormality can manifest itself without the treatment.

Long before these experiments, however, embryologists knew of natural cases of this effect. Take, for example, the callouses on the feet of human embryos. These could have arisen in a similar Darwinian fashion from ancestors who originally acquired callouses only after birth. Those who developed them more quickly might have had a selective advantage. Their offspring would not have to stamp around so much to get callouses. By assimilation, we have reached the point where expectant mothers carry babes with calloused feet.

Genetic assimilation is repeatedly demonstrable; but yields to Darwinian explanation. Another experiment reported recently by Gorczynski and Steele (1981) would seem less amenable to such explanation.

Steele's discovery is of an allegedly Lamarckian effect in the immune system of mice. As is well known, foreign particles called antigens can be recognised and combated in higher animals by antibodies produced by the immune system. There are approximately 10^7 different antibodies. But there is DNA for only about 10^6 genes in the genome, a large part of which either is non-functional or does not code for proteins. The DNA required for the production of antibodies must consequently arise somatically. That is, it must arise in the cells which produce the antibodies and be absent from the germ cells. And indeed recombination and mutation, working on a few DNA molecules to start with, can produce a great many new ones ready for immunological purposes when they are needed. During a particular embryonic phase the immune system 'gets to know' the antigenic properties of the surrounding body, in order to exempt them from attack. (Failure to do so results in some serious diseases such as rheumatoid arthritis, pernicious anaemia and others.) During this time one can 'trick' the immune system and make it tolerant to foreign antigens mistakenly taken for its own. This Steele and his colleagues did with their mice. Apparently the tolerance to foreign antigens thus induced was transmitted as far as the second generation, without descendants having been made tolerant during their own critical phases.

Subsequent attempts to repeat these experiments have failed (Brent *et al*., 1981; McLaren *et al*., 1981; Medawar *et al*., 1982). But should they nevertheless hold up (which seems unlikely at present), and should they really call for a Lamarckian interpretation, this would mean at most that for one system – the immune system – natural selection was not the only mechanism of adaptive evolution.

Given present-day knowledge in molecular biology, what possible mechanism of information transfer could subserve Lamarckian inheritance? Almost a century ago Weismann (1885) propounded the doctrine that soma and germ line were separate systems within any sexually reproducing organisms. What Lamarckism requires is an information flow from soma to germ line over and above the reverse flow that takes place in ontogenesis as the organism 'expresses' somatically the information provided by the germ line cells that give rise to it. Molecular biology tells us of an information flow from DNA to the messenger molecule RNA which enters the cell cytoplasm to direct the synthesis of the protein for which that DNA codes. Crick's (1958) so-called central dogma of molecular biology denies the possibility of any backward effect, that is, of information flow from protein end products to nuclear DNA. This, however, does not amount to the denial of any form of Lamarckism whatever. For there is still the Lamarckian possibility of information flow from RNA to nuclear DNA. Without going into the technical details, it is this possibility which Steele would have to exploit if his experiments are to sustain a Lamarckian interpretation.

16 ORIGIN OF LIFE

One of the first experiments on the origin of life was performed by Miller in 1953 (Miller and Orgel, 1973). It concerned the 'spontaneous' formation of prebiotic molecules. Before then biogenesis had been a topic of wild speculation. Following Lerner (1968) we may describe the five main theories as follows:

1 Divine creation

In its literal biblical version, this theory today appears grotesque. But one should not forget the serious-minded attempts by scientists in the last century to work out whether the world could have been created in six days or whether Noah's ark had room for representatives of all animal and plant species. More recent divine creationist theories that have evolved through dialectical osmosis with evolutionary theory itself (such as Teilhard's) have been too vague and mystical to represent any useful advance on the latter. Medawar's broadside on Teilhard's theory ended by dismissing it as a 'bag of tricks'.

2 Omnipresent life
That life has always existed on earth from its very beginning contradicts geological knowledge. The fossil record reveals no life forms from 5 billion years ago, and shows that simple forms arose only about 3 – 4 billion years ago.

3 Life from outer space
Other star systems or galaxies might have exported simple life forms to earth, and these could have 'seeded' the evolutionary process here. Although not wholly impossible, this theory appears highly improbable. First, no trace of such life has been found in meteorite and comet remains that have reached earth; and second, there would still be the problem of how life began, wherever it came from in outer space. It has even been suggested that the earth is a planetary laboratory for alien biologists! Presumably we shall never know the outcome. The theory postulating this experiment hardly appears testable by experiments within that experiment, especially if the aliens have designed it that way!

4 Continual creation on earth
The reigning conception after Aristotle had it that besides ordinary conceptions there were spontaneous events of creation ensuring the proliferation of new species. Insects and worms were held most often to arise in this manner. Francesco Redit (1626–98) was the first to show that no flies would emerge from meat unless fly eggs were there beforehand. Then Leeuwenhoek's discovery of micro-organisms gave this theory a new impetus. It finally fell into discredit when Pasteur showed that no form of life would develop in any sufficiently sterile nutritive medium.

5 Life has arisen just once, or at most a few times
Dating such events as having occurred between 3 and 4 million years ago gives us a 'creation' hypothesis that is well supported both experimentally and theoretically by neighbouring disciplines. Spectroscopic analysis of other planetary atmospheres gives us a good view of the earth's atmosphere at that time. It was probably a reducing atmosphere, whose main constituents were methane,

ammonia, water and hydrogen (but see Halvorson and Van Holde, 1981, for alternative theories). Miller simulated lightning flashes in such a mixture of gases, and thereby obtained complex molecules that included the amino acids glycine and alanine. These are two of the twenty or so amino acids that are the building blocks of proteins, which in turn are the basic organic material of which organisms are composed. Many variations of Miller's experiment since have shown that all the chemical preconditions for the creation of the first living entity were met in the primal atmosphere. The important building blocks of life, namely polypeptides (small protein molecules) and nucleotides (the basic hereditary substances), have been spontaneously synthesised in these experiments.

It is, of course, a huge step from the presence of these building blocks to the emergence of the first functioning cell. It has not been accomplished under laboratory conditions, but the presumed stages of transition do not involve anything incompatible with present physicochemical laws. Nor do they force us to postulate any new forces for their explanation. Fox (1973) has succeeded in creating so-called microspheres – cell-like forms consisting of protein molecules, that can divide and that have a kind of cell wall. The experimental conditions that produced them were designed to be like those that must have prevailed in the vicinity of an active volcano. A mixture of amino acids was heated to between 150 and 200°C resulting in their polycondensation. When the products were placed in a watery solution, the astonishing bacterium-like microspheres resulted. They even have a primitive metabolism, are photochemically active, and can be stained in the way bacteria can. To get from microspheres to bacteria, one would have to introduce nucleic acids containing the information concerning the amino acid sequence of the proteins making up the microsphere. Because of the specific possibilities of chemical bindings involved, this poses some brainwracking problems to the biochemist, but none that are theoretically insuperable. Indeed, Eigen and Winkler-Oswatitsch (1981) believe that the characteristic association between proteins and nucleic acids occurred even at a precellular stage. His theory of the origin of life shows how the Darwinian principle of natural selection operates even at the 'purely chemical' stage. The probability that a cell would eventually be 'created' by natural processes thereby increases considerably. No longer does one need to postulate a purely accidental collocation of all those constituents that go to make up a functioning cell. The interplay of chance and necessity so characteristic of modern Darwinism begins in the primal soup with the genesis of self-reproducing molecules.

Simple RNA molecules have been shown to be able to assemble and replicate in cell-free systems containing lead or zinc ions as catalysers. Each self-assembled molecule can serve as a template for the production of complementary strands. This replication would occur at a rate proportional to the concentration of the molecules – so the growth rate would be exponential. Furthermore, more varied molecules would be created by errors in replication.

These conditions are of course ideal for a selective process. The most stable molecules that replicated fastest and most reliably would use up most of the raw material when making copies of themselves, and their number would swell disproportionately within the population of molecules. Eigen calculates that their maximal length would be 100 nucleotides, this being the reciprocal of the error rate of 10^{-2} in replication. If any longer than this, the error rate would guarantee the eventual attrition of the parental sequence.

To get from this stage of pre-biological evolution to the co-operative interaction of nucleic acids and proteins, which seems necessary for the development of living systems, Eigen and his colleagues (1981) hypothesise the so-called hypercycle. A hypercycle involves autocatalytic processes allowing 'positive' feedback. Assume that a certain RNA sequence (I_1) codes for the enzyme (E_1) (see Figure 1.1). This in turn catalyses the replication of another RNA sequence (I_2) which codes for a second enzyme (E_2). Under these conditions the two RNA sequences are forced to co-operate because they depend on each other. Hypercycles might grow in complexity and efficiency, and if enclosed by a primitive cell wall might become the first 'organism'.

Whether things really took this route (H. Kuhn and Waser (1981) suggest a different one) is not very important. What is important is that it is physico-chemically plausible. The first steps in the evolution of living systems contain no processes that are inexplicable in principle. Eigen's work shows how selective processes could have started in prebiotic times. The argument from 'low probability of emergence of life' therefore suffers another serious blow.

Kuhn and Waser, without hypothesising hypercycles, calculate that in 10^8 years chemical evolution could produce the 1,000 odd proteins in a bacterium. Evolution has had 10^9 years. Give or take an order of magnitude, life comes within chemical compass quite plausibly.

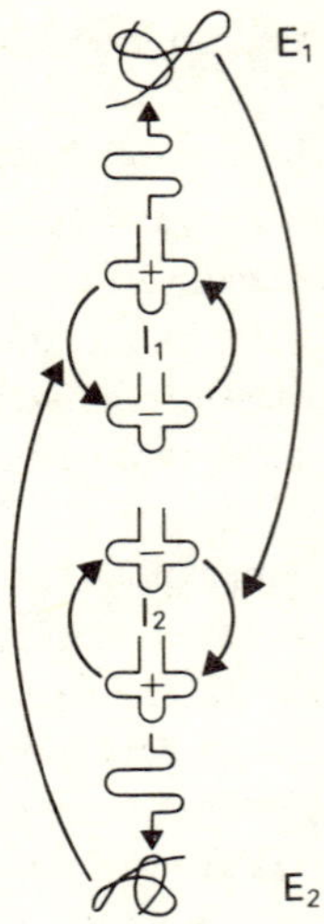

Figure 1.1

Eigen *et al.* (1981) demonstrated more than just a possible scenario for the origin of life. They even calculated the most probable nucleotide sequence of the first RNA-molecule. Their work shows that Darwin's law of natural selection cannot be used as an argument for the autonomy of biology. Many authors (e.g. Ayala, 1974) have claimed that physics has no equivalent of a Darwinian selection process and that biology therefore cannot be reduced to physics. Although we agree that such reduction is not possible (see ch. III, section 2), we disagree that natural selection is the reason for this. Eigen showed that Darwinian selection is no special mark of the living world setting it off from the physico-chemical.

The basic goal of all theories of biogenesis that make modern scientific sense is to interpolate as many plausible intermediate stages as are needed to explain the transition from the primal soup to the first bacterium-like cell. As we have indicated, it is highly unlikely that all the constituents of the first cell would have come together with no intermediate transitions.

Presumably the weakest point in Eigen's model of the emergence of life is the transition from a non-functional connection between proteins and nucleic acids to one in which nucleic acids contain in their base sequences the information for the amino acid sequences in proteins.

But there are hypothetical suggestions as to how this too could have occurred. t-RNA is a nucleic acid found in every cell. Relatively short nucleic acid molecules very similar to present-day t-RNA could have tacked themselves at specific sites on to longer RNA molecules (called messenger-RNA) 'nose first', dragging along specific amino acids on their 'tails'. Thus the messenger RNA molecule would be like the leg of a ladder, the t-RNA-like molecules attached thereto like floating rungs, with only some way needed of joining the trailed amino acids at the ends of those rungs together, to form a detachable protein molecule.

Once self-replicating entities have come into existence, then, as Darwin recognised, further 'creations' become highly improbable. The existing systems will prey on any underdeveloped newcomers. Parallel development can take place only in inaccessible niches. These, however, eventually have to open up to outside competitors. And it can hardly ever happen that the natives win the competition. For, in all living creatures, from the most primitive bacteria to man, one finds exactly the same chemical foundations of life (with the possible exception of mitochondria). The genetic code is universal, pointing strongly to a unique ancestral hypercycle. The intervening evolution of eukaryotic cells, multicellular organisms, fish, mammals, primates and men, is thoroughly described in any modern textbook on biology. It is worth mentioning here that biochemical evidence indicates that man's common ancestor with chimpanzees and gorillas is nowhere near as far in the past as is commonly assumed. Biochemists now suggest that the period of descent in question is only four million years (Cherfas and Gribbin, 1981).

17 THE THERMODYNAMICS OF LIFE

A closed system is one which exchanges no energy or matter with its surroundings; and the first law of thermodynamics says that the energy contained in such a system is constant. More relevant to our considerations is the second law. It says that all forms of energy can be converted into one another but that thermal energy (heat) can never be converted entirely into any other of the free forms of energy. The result is a constant increase of thermal energy at the expense of all the other forms. Ultimately this leads to the so-called heat death of any closed system. The entropy of a system in a given state is a measure of the probability of its being in that state; and in heat death entropy is maximised. The second law, or law of entropy, as it is also called, has given rise to two debates in the philosophy of science. One, which we shall not go into, concerns the nature of time. The other concerns the compatibility of evolution with thermodynamics. This problem now appears to have been solved. Anticipating: there is nothing in the laws of biology that conflicts with the law of entropy. The biosphere is anything but a closed system since it constantly exchanges matter and energy with its surroundings. Nor is the earth a closed system, since it absorbs massive amounts of solar energy. Only the solar system can be approximated as a closed system.

How did apparent contradiction arise? Superficially, evolution appears to proceed by successive reproductions from relatively unstructured to ever more complex entities. This increase in complexity and consequently of information means that entropy is decreasing in the containing system.

Thus *if* that system could be regarded as closed, the second law would be violated. Following C. F. v. Weizsäcker (1974), we shall consider the four main attempts to accept or avoid contradiction.

1 Vitalists maintained that entropy does decrease during evolution and that biology consequently contradicts physics. Naturally scientists found this quite unsatisfactory as it amounted to no more than admission of defeat.

2 Some evolutionary biologists thought they could avoid the problem by denying any increase in organic complexity during evolution. Indeed, it is practically impossible to measure the complexity of a system (but see the discussion of this problem in Peacocke (1983)). It is also practically impossible to say how many of the chosen units of information – for complexity is by and large synonymous with information required for construction – separate a bacterium from a chimpanzee. Objectively minded biologists might have sought some quantification of complexity as a reaction against the mysticism of writers such as Teilhard. But there are some qualitative arguments also for the view that functional and morphological complexity of oganisms increases during evolution. The reason why it is difficult to quantify the increase is that the current concept of information as defined in information theory is not suited to this purpose. First, this is because the notion of information we are seeking is one that

will permit one to distinguish between the sense and nonsense of two otherwise equally complex systems, one of which 'makes more sense' than the other against a given background. Second, because it is not at all an easy matter precisely to define the concepts of energy, information and probability and to determine the relations among them. One requires very little information to describe highly probable states of affairs – that is, states of high entropy. Probable states of affairs always contain a uniform distribution of their constituents whose description requires little information. This is intuitively graspable in the case of a crystal. It is a physical fact that the more regular a crystal is, the more entropic it is. But now it is also clear that the more regular crystals are, the easier they are to describe. This example serves also to distinguish being ordered from being complex. Crystals are highly ordered but not at all complex. Every departure from regularity entails a decrease in entropy and an increase in the information required to describe the system. The interested reader is referred to Kiefer (1979) and several authors in Weizsäcker (1974).

3 Many physicists believe that entropy does indeed decrease during evolution, but that this decrease is compensated for by an increase in another part of the system. A not altogether serious but nevertheless instructive example that may clarify this idea at least for German readers has been given by Stegmüller (1975):

> A scientist working with concentration creates a large amount of negentropy (that is he reduces the entropy); the resulting decrease in order that compensates for this in his surroundings is clear for all to see in his untidy desk. By contrast a bureaucrat – for example one busy translating a local council regulation into a university statute – creates little negative entropy. This is shown by the fact that his desk is in meticulous order. (our translation)

The biosphere is not in thermodynamic equilibrium; for any organism disintegrates upon isolation. A crystal, however, is stable even in isolation. The biosphere's position far from equilibrium is maintained by a constant flow of energy and matter through it. It is in a state of *dynamic* as opposed to *static* equilibrium.

Open systems far from equilibrium involve irreversible processes which have been investigated recently by Prigogine (1979; 1980; and

Stengers, 1981) and his school. The essential results of their mathematically demanding work can be expressed simply as follows. Open systems in a constant energy flow reduce their entropy production to a minimum. A didactically and aesthetically beautiful example are the so-called Benard cells that arise in thin sheets of dense liquids when these are heated from below. Regular honeycomb patterns are formed by convection streams. As with all such dissipative structures (where by 'dissipation' one means the transformation of free energy into useless thermal energy) the hexagons disintegrate as soon as one interrupts the energy flow. Prigogine believes he has discovered the secret behind any capacity for self-organisation far from equilibrium, not only at the physical level but also at the biological, psychological and sociological levels. An important point about dissipative structures as far as biology is concerned is that they may remain stable even when the energy flow creating them has been removed, provided that chemical bindings maintain them. The thermodynamics of irreversible processes holds out most promise in biology in the understanding of morphogenesis. It is known that morphogenetic processes, processes giving rise to three-dimensional structures, take place in a field of chemical or physical gradients. Without these, living cells could not orientate themselves and morphogenesis would be impossible. Morphogenesis is nothing but the spatio-temporally organised summoning up and execution of genetic instructions. Genetic information by itself, unconnected to given structures that make orientation possible, would succeed in creating only cancer-like growth.

These ingredients are possibly set up according to a principle of self-organisation and may not be laid down in the genome. Very plausible models of these have been developed by Gierer (1974).

The biosphere is an open system in which entropy is declining; and this decline is partly set off by an increase in the entropy of the physical environment and is paid for in the main by solar energy. If, therefore, we regard the solar system as closed, then it is the sun's thermonuclear processes that compensate for entropic decline in the biosphere. This will, of course, last several million years during which time the entropy here on the earth's surface could decrease further. The sun's energy, of course, was inherited from the big bang (if it ever took place). So the fact that entropy is decreasing in our biosphere is to be attributed to those processes following the big bang which prevented energy and matter from being uniformly distributed. Were these also dissipative processes?

4 The last of the four possible ways of dealing with the apparent contradiction between evolution and the second law of thermodynamics is the one favoured by Weizsäcker himself. He does not believe in any contradiction, as high entropy has too hastily been equated with unstructured homogeneity. As the example of a crystalline body shows, a highly structured object (which is not to say a complex one) can possess more entropy than a homogeneous mixture of its constituents. Temperatures on earth in comparison to those possible – for example 10^7 °C in the interior of the sun – lie near the 'absolute zero' of –273°C. Under these conditions, according to Weizsäcker, heat death is not a brew, but a collection of complex skeletons. Thus the evolution of ever more organised systems does not contradict the law of entropy. This development follows from the law of entropy itself in applications to closed systems. If we have understood Weizsäcker correctly, then, one does not need to seek a solution to the problem in the thermodynamics of open systems.

Weizsäcker's solution is problematic in so far as it is not clear whether he is not here confusing order with complexity. A crystal is an object with a high degree of order but a low degree of complexity. Decrease of entropy follows only from increase in complexity, not of order. In any case, should Weizsäcker be guilty of this confusion, the resolution of our problem still follows from all that we said under heading 3. There we saw a simple solution in terms of open systems far from equilibrium, of which the earth is an example.

18 THE SCIENTIFIC STATUS OF EVOLUTIONARY THEORY

Recently, debate has intensified over the scientific status of neo-Darwinian theory. The *dramatis personae* and the plot are at times bizarre. Creationists are seizing with glee upon some maladroit misgivings on the part of philosophically naive paleontologists and museum keepers, over whether evolutionary theory passes muster as a genuinely scientific discipline. The creationists, of course, somehow manage to see around the beam in their own all-seeing eye. The philosophically naive scientists invoke the name and support of the philosopher of science Popper. Nearly all of Popper's scientific experience has been in the physical sciences, and the philosophy of biology has recently got up its own head of steam precisely because of the inadequate methodological conceptions available within a tradition that has concerned itself almost exclusively with the physical sciences: with such matters as space and time, causality, matter, energy, measurement, the observational/theoretical distinction and so on – with hardly a word for the special problems posed by a scientific study of living organisms in all their uniqueness and complexity. Finally, recent realignment within the Darwinian camp over the significance of the fossil record, and the question whether speciation was gradual or punctualist (in the sense discussed above) has been misinterpreted by popularisers of science as evidence that the theoretical edifice is cracking.

We hope by the end of this section to have put all these matters into a clearer perspective, through careful consideration of particular criticisms that have been made and that are worth considering. To this end, we shall also be mentioning several examples of 'laws' of evolutionary theory and discussing how they operate within the overall framework. But first it is worth giving a resumé of the desiderata, widely agreed upon by philosophers of science, for a given body of assertions to constitute a scientific theory. We shall see that the well-known insistence on falsifiability is just one among many, all of which must be considered both when choosing between rival scientific theories, and when deciding whether to reject a purported theory as non-scientific, or 'metaphysical'. Our conclusion will be that on all scores, the synthetic theory of evolution is scientific, and indeed has the field to itself at present.

Certain desiderata are so obvious as hardly to be worth mentioning. The theory should at least be formulated in a language that is understood, and should be internally consistent. People should be able to agree on application of at least some of its terms, such applications providing the 'connection with reality' or world of experience which, in its essentials, is what the theory is trying to describe. This does not imply that we should be able to make out a clear distinction between 'observational' and 'theoretical' terms: this distinction has come under heavy attack in recent philosophy of science, but this will not concern us here.

Popper (1959) is justly famous for his insistence that the theory should be falsifiable by experience or experiment. Otherwise, if no experience whatever could be regarded as falsifying it, it can be no more than metaphysical dogma. This in turn means that the theorist must be willing to state in advance what sort of evidence he would count as falsifying his theory. Obviously data are very often evaluated in the light of theory. The absence of a clear-cut distinction between the observational and the theoretical means that some phenomena or happenings will be described at times in 'theory laden' terms. But theoretical redescription of evidence must have its limits. There must be an end, that one can specify in advance, to the juggling with labels and massaging of data than can be undertaken in any attempt to get out of a sticky experimental patch.

This having been said, however, we should not lose sight of the fact – testified to by many historical cases – that good scientific theories have simply tolerated anomalies (even if only because there have been no good alternative theories), while fully admitting that matters could not be redescribed in any way that would make recalcitrant experience consonant with theoretical pre- or postdiction. As Quine (1953) has so often described matters here, with reference to Neurath's image of the floating raft being rebuilt plank by plank while at sea: our theories face the tribunal of experiences as articulated wholes. Scientific advance to a large extent consists in distributing truth values within a network of sentences, with the more 'observational' ones at the periphery being more responsive to impinging experience. This is at odds with that puritanical version of Popperianism that demands the rejection of any theory should any one of its predictions be falsified by experience. As a description of how scientists actually proceed, this is way off-beam (Feyerabend, 1975). As prescription it raises the question of just what sort of rational agents it will help. We certainly have grave suspicions that it will *not* help rational agents whose very own reason has been produced by a process of adaptations much like the rebuilding of the raft. Pristine Popperianism sees theoretical advance as consisting only of the survival of epistemic 'hopeful monsters', with no intervening periods of intra-specific adaptation. Seeing hypothesis formation and testing as analogous to mutation and selection is unlikely to lend support to the extreme form of Popperianism we are discussing.

Any theory worth its salt should make predictions or postdictions not yet contained in the evidence on the basis of which the theory was erected. Theory must 'go beyond' data. The boldest way to do this, of course, is by containing laws of universal form, generalisations maintained to hold for all times and for all places, and for all entities of such-and-such natural kinds. This, indeed, may turn out just to be a special case of simplicity. As Sober (1975) has analysed the latter, simplicity consists in the ability to answer questions about well-defined states of affairs while yet appealing to as little extra information about the set-up as is possible. If a simple, universally formulated theory withstands rigorous testing by experiment and new observations, the theorist may well begin to believe that he has, in Quine's phrase, 'limned the ultimate traits of reality'. And, of course, the bigger the chunk of reality limned the better. A theory that integrates pre-existing tracts of knowledge, subsuming them into an overarching scheme with a uniform stock of concepts and new laws drawing together those domains of experience perhaps previously thought separate or incommensurable – such theories are the stuff scientific dreams are made of. Moreover, it is important to note that this subsumption of the disparate can count strongly in favour of a theory even if it does not deliver itself of many cast-iron predictions. In such a case one speaks of inference to the best explanation, or abduction as Peirce (1958) called it. Finally, it is a mistake, perhaps induced by the paradigm of Newtonian dynamics, to think that all theoretical laws must be unrestrictedly universal and mention only determinate properties. Not every law says that every F is (most certainly, definitely) G. Indeed, with the advent of quantum physics, in statistical sciences we now see such apodeictic laws as the exception rather than the rule. Hempel (1962; 1965) and other philosophers of science have made it abundantly clear that perfectly good science can deal in probabilistic or statistical laws, whether or not these in turn are arrived at by inductive methods or by the hypothetico-deductive method recommended by Popper, namely the method of bold conjecture and assiduous attempt at subsequent refutation. When applying such laws experimentally, moreover, a great deal must be left to the practitioner's craft in his discipline, to his ability to judge whether the '*ceteris paribus*' clauses built into any scientific law are indeed met in the experiment at hand, especially when it appears to hold out the prospect of refutation.

We now pass on to consider how the synthetic theory of evolution fares according to all these desiderata. Popper has, in several of his writings (e.g. 1966), criticised evolutionary theory on two grounds. (The fact that in his autobiography *The Unended Quest* (1974) he regards the prediction of gradualness as the only prediction of the theory should not prevent us from looking at his objections more closely.) First, evolutionary theory is allegedly tautological, circular or empty. The fittest organisms are defined as those that survive (to reproductive age), so the survival of the fittest amounts to no more than the survival of those that survive. Second, the evolutionary process is singular, unique and unrepeatable. Thus the statement that related biological forms are of common ancestry is a singular historical statement, neither a law nor law-like. At the very best evolution proceeds in accordance with laws (presumably those of physics and chemistry), but the statement in question is not itself a law.

There is much to sort out here. First, let us be clear about the object of these criticisms. What, in a nutshell, *is* the theory of evolution by natural selection? The historical statement about ancestry given above is hardly the whole theory. It is rather a portmanteau encapsulation of observed data, like Kepler's laws of planetary orbits, which were based on Tycho Brahe's painstaking observations. Just as Kepler's laws called for theoretical explanation – and eventually received it in Newton's theory of motion and gravitation – so the historical statement that related living forms have common ancestry calls for theoretical explanation; and that theory is what we have to characterise. The usual wholly inadequate formulation of the theory, in the form not even of a proposition – 'the survival of the fittest' – entirely omits the main insights on which the theory is based, even though it may be the historical reason why the allegation of tautology has arisen. It is a slogan and no more; and the greater is the pity that Popper should not have acquainted himself more thoroughly with the details of the theory he was criticising. Happily, in later work he has recanted his criticisms; but such was his stature that they have had a lasting effect, which it is our concern to undo. It is worth digressing to point out that no one appears to have replied to the charge based on the slogan, with the obvious defence that fitness may admit of independent measurement. Methods may be developed that would free us from *ex post facto* analysis. According to the theory of sexual selection, sexual partners assess each other's general fitness during courtship. Could not their implicit recognitional modes be converted into explicit scientific criteria? We are after a dispositional property of the organism, one that incorporates probability estimates. Exactly the same holds for the dispositional property of a die to land on the six. We might adopt here the wait and see method, basing estimates of probability on long run frequencies; or, on the other hand, we might investigate the die for homogeneity and geometrical regularity in an attempt to determine that probability independently. So why not admit the possibility of doing something analogous as far as fitness is concerned?

We were at pains in section 1 to set out three necessary and sufficient conditions for evolution to take place in a population of self-replicating entities. We pointed out how there is ample empirical evidence that the three conditions are satisfied in our world; and that the overall conditional claim, to the effect that *if* these conditions hold *then* evolution takes place, was a logical truth. Note that nowhere in the antecedents or the consequent of this conditional do we find the phrase 'survival of the fittest'. In exactly similar fashion, one will not find the heuristic aphorism 'every action has an equal and opposite reaction' receiving mathematical expression in the axiomatised version of Newtonian dynamics. It is a gloss that drops out of the vector-algebraic formalism.

Where, now, does the alleged tautology reside? We have already explicitly conceded that the core conditional of evolutionary theory is a logical truth. But that matters not a jot. The conditional says roughly that once a 'life-permitting situation' arises (that is, one for which the three conditions hold), then evolution follows as a matter of logical necessity. To see this as scientific bowdlerisation is absurd. One could equally well charge Newtonian dynamics and gravitational theory with vacuousness, for saying that once an 'orbit-permitting situation' arises then it follows as a matter of logical necessity that orbits will be elliptical. There is a possible objection to this analogy with Newtonian dynamics. It may be objected that built into the specification of an 'orbit-permitting situation' is the condition that the gravitational force obey the inverse square law; and *that* it does is an important statement of the theory to be tested. Thus what is being tested in the Newtonian case lies in the antecedent of the logically true conditional, whose very logical truth is what allows one to explain the consequent by invoking the antecedent(s). Under pressure from this objection, our response would be to confer greater theoretical status on the antecedent condition in the evolutionary case that says that phenotypic traits are heritable. *That* they are is, no matter how reasonable, nevertheless a *theoretical* conjecture on a par with the Newtonian one to the effect that there are gravitational forces at all. Now, in the Newtonian case one can go on to ask 'What *kind* of force – how large is it, and what does it depend on in general?' The answer provided by Newton was, of course, that the gravitational force between two bodies depends in the well-known way on their masses and the distance between them:

$$F = \frac{GM1M2}{R^2}$$

Similarly, in the case of evolutionary theory, one could go on to ask 'What kind of heritability – how strong is it, and what does it depend on in general?' The answer provided by Darwin, was, of course, wholly inadequate, although the modern reader will see sufficient material in the early chapters of the *Origin of Species* to wonder why Darwin didn't fall off the logical log through the barn door of particulate inheritance. But the founders of the synthetic theory have since made good this shortcoming. It is as if Newton himself had only got as far as his laws of motion, and the thought that heavenly bodies were somehow attracted to each other, but only God knew how; and someone like Laplace or Hamilton had to come along with the inverse square law to get the synthetic theory of planetary orbits going. What Darwin's successors importantly did was to supply the missing conjecture as to the exact nature of the hereditary mechanism. The synthetic theory can now be broken down into the following components:

1 The mass of observational evidence supporting the claims that there is phenotypic variation; that at least some of this is heritable; that organisms have different numbers of offspring; and that the distribution of phenotypes in populations and the number and kind of species changes over time.

2 The 'core conditional' alerting one to the necessary connection between variation, heritability and differential reproduction, on the one hand, and evolution (change in distribution of phenotypes) on the other.

3 The postulation of natural selection working on random genetic mutation as the one and only mechanism of evolutionary change.

Where now can the falsificationist find fault in the synthetic theory? Let us look at the empirical status – not as true or false, but as falsifiable or non-falsifiable by any possible observation or experience – of the claims in (1) *Phenotypic variation*? Look around you. You either see it or you don't. It's there. *Heritability*? This time the answer is more subtle. Certainly, it seems that traits run in families, that there is not just a random correlation between features in the parents and corresponding ones in their offspring. Perhaps, via the conditional in (2), evolution itself is the ultimate evidence for traits' being heritable! (Just as planetary orbits provide good 'ultimate' evidence that the sun exerts some sort of attractive force on the planets.) But we also have a great deal of evidence that offspring are like their parents even when brought up away from them and in very different conditions from them. *Differential reproduction*? Again, look around you and keep count!

Finally, *evolution*, both micro and macro? Here the evidence does not strike the uninformed eye until it is trained to look for it. The 'singular, historical' statement that evolution has produced today's species from ancestral ones, from perhaps one or a few original species, by means of *relatively* gradual morphological change, is obviously falsifiable by the fossil record. Many a biologist has expressed preparedness to jettison this portmanteau claim should we discover inversions of the 'tree of life'. We must, of course, always bear in mind the factors determining what bones are laid to rest where. Animals migrate, invade new niches, burrow, and might speciate in very small numbers leaving no trace intact for us to dig up and evaluate later. And indeed, such imperfections of the geological record as there are can be quite adequately explained by considerations such as these. But, of course, when such considerations are adduced (such as, say, the claim that an apparent 'jump' in organic complexity of fossils at a particular site is due to the arrival of a sophisticated breed of migrants from elsewhere) then they in turn must be enmeshed in a factual basis; and such is the mosaic of conjecture as to datable occurrences that paleontologists and geologists are doing their utmost to piece together. Interestingly enough, Darwin himself did not set too much store by the fossil record, which he acknowledged as incomplete. But it is worth pointing out that attending to the morphological continuities possible or plausible as one pieces together the mosaic referred to was precisely what enabled scientists to suspect that the skull of 'Piltdown man' was a fake (Root-Bernstein, 1981). As for further evidence of evolution: the reader of the *Origin of Species* will find himself almost prepared to grant Darwin his whole case after only the first chapter containing his acute observations on the breeding of dogs, sheep, pigeons and other domestic species, by both conscious and unconscious selection on the part of man. There is even evidence today of speciation through artificial selection. As Ridley (1981) points out, artificial selection is experimental evolution; and we have seen such evolution of a new species in *Primula kewensis* from *Primula verticillata* and *Primula floribunda*. Moreover, the natural species of mint *Galeopsis tetrahit* has been recreated artifically by crossing *Galeopsis pubescens* and *Galeopsis speciosa*. To end this paragraph on the facticity of evolution, we must point out that the possibility (suggested by the punctualists) of rapid speciation followed by stasis, in no way undermines any essential tenet of Darwinism. If the populariser wishes to sum up the theory of the *Origin* with Latin phrases such as 'natura non fecit salta' and then wishes to interpret periods of tens of thousands of years as literal points of discontinuity, we shall simply allow him his semantic flights of fancy. We can expect major redistributions of emphasis, reinterpretation and reassessment of evidence in our inferential quest to the best explanation of what lies inert underground and alert above it, without capsizing or sinking our Darwinian raft, the synthetic theory. This is not to say, however, that anomalies could not arise which would force us overboard. We just have so much positive evidence that it now provides a great deal of room for manoeuvre when difficulties arise. Bearing in mind that evidence bears holistically, and that *ceteris paribus* clauses can speak volumes is not the same thing as clinging to untestable metaphysical dogma.

By far the most interesting and crucial claim is (3), that the mechanism of evolution is natural selection based on random variation. The existence half of this claim is hardly open to doubt. Observations in the wild by geneticists and others, and knowledge of the molecular chemistry of the genetic system, disclose both selection and mutation. The uniqueness half of the claim remains a working hypothesis. Lamarckian inheritance might be a competitor with an inside lane one day. (Chance events, too, might come to be accorded a more determinate status.) That will, however, be like the Newtonian discovering that besides gravitational forces obeying the inverse square law there were also electromagnetic forces obeying their own ones. Allowing them into the picture, however, does not impugn the validity of the basic laws of motion. These say how bodies move when forces and masses are known. That there may be forces whose direction and magnitude depend not on mass and distance (such as the gravitational) does not falsify the central Newtonian claim that force is the rate of change of momentum. Likewise, admitting Lamarckian inheritance and the new selective forces it entails will not alter the theoretical core given by (1) and (2) above.

We have thus far been concerned with the core of the theory of evolution, trying to draw a tight enough analogy between it and the Newtonian theory of dynamics to establish its credentials as an empirically falsifiable theory. It will also be useful to look now at some further generalisations, laws and rules of evolutionary theory in application to specific areas. Continuing the analogy, these will correspond to such laws as Hooke's Law for springs, the law of friction and so on, the auxiliary theses brought to bear in special domains of application of Newton's theory. We can enquire further whether these more specific evolutionary laws are unrestrictedly universal in scope, deterministic or statistical, and whether they can be applied so as to yield testable predictions. In other words, are they scientific laws at all?

First, we have the laws or rules of the early evolutionists: Bergmann's rule, Dollo's law and Haeckel's law.

Bergmann's rule is that closely related species of one genus or family tend to be larger the further north their habitat is from the equator. One expects this because a lower ratio of surface to volume keeps body heat in. On closer inspection, however, it turns out that the rule holds in only about 70 per cent of cases. This leads some to suggest that it is not a law at all – presumably because a law is not supposed to have exceptions. But the exceptions can frequently be explained. For example, a burrowing rodent is better off with a small body in the harder, colder ground in polar regions. Bergmann's rule clearly has a hefty *ceteris paribus* clause built into its applications.

Dollo's law states that any path that the evolution of some trait or organism might have taken can never be reversed. On the molecular level there is an exception to this, because any point mutation can back-mutate and thereby restore the former state. As we have pointed out several times already, however, major changes in complex structures can not have been brought about by single point mutations. The more mutations, recombinations and chromosomal re-arrangements are involved in evolutionary change, the less probable its reversal becomes. For example, whales are descended from formerly land-living mammals. Mammals themselves, of course, descend from fish. When whales returned to the water they did not become fish again. They retained their lungs, and their swimming movements are not in the horizontal plane (as is the case with fish) but in the vertical (as is the case with running mammals). Whether propulsive forces are generated by horizontal or vertical movements may not be of great importance; but the absence of gills in a creature living in water surely is. The general thrust of Dollo's law is underwritten by what we now know of the genetic mechanisms underlying evolution. The synthetic theory, as Ridley (1981) points out, is uniquely able to explain not only the presence of design, but also the absence of design (as is the case with vestigial organs). Absence of present design is most often obsolescence of past design. If the past trait had not been selected for, then it might nevertheless still have arisen by selection for other traits, whose co-presence produced it as a relatively harmless excrescence.

The problem of judging the adaptiveness or otherwise of an identifiable trait is not to be underestimated. Some biologists, such as Gould and Lewontin (1979), have argued recently that evolutionary theorists have been far too ready to see every trait of an organism as having adaptive significance, and calling for evolutionary explanation. No doubt knowledge of past environments is a crucial prerequisite for determining whether there might have been selection *for* a trait, rather than for others which in turn produced the trait in question as a by-product. An analogy might help to bring these thoughts into clearer focus. Suppose one is erecting a structure like a Bedouin tent, with several poles of different length pitched in different directions to prop the canvas up. The relative positions of the poles and pegs determine the way the canvas will form sagging or saddle-shaped areas. The surface geometry of the canvas is a byproduct of the pole and peg positions. Now it *might* be the case that pole lengths are determined only by quirks of the constitution of timber – that the strongest poles come in lengths that are multiples of six feet, with strength increasing proportionately with *those* lengths, but not with intermediate ones. The overall stresses that the canvas is to place on the poles could determine what lengths are placed where; and this in turn could be relatively independent of how streamlined the tent surface may be in the face of winds of different strengths and from different directions. *That* the tent has such-and-such a shape would therefore *not* be best explained by appeal to the aerodynamic properties of the tent surface, but rather by appeal to its being a by-product of pole and peg placements that have been determined by other optimising constraints. The question is thus always: which are the basic adaptive traits which 'generate' the rest?

Another analogy, bringing us back to the realm of Newtonian mechanics, is introduced by the similar question: which are the basic co-ordinates to choose for the best vector description of a physical system of masses in motion? In the case of planetary orbits, polar co-ordinates are by far the best, because one can treat the sun as the centre and gravitational attraction as radial. In general, inapposite choices of co-ordinate system in mechanics leads one to have to postulate 'unnatural' forces (such as coriolis forces with spinning bodies) to account for the motions observed. This is very much like latching on to the wrong trait in the evolutionary case and looking for selective forces to explain its evolution as part of adaptive design for the organism concerned. That some evolutionists make this sort of mistake, however, in no way detracts from evolutionary science as such. To press the analogy further, consider Einsteinian mechanics. If an Einsteinian physicist insists on keeping Euclidian geometry, he can still do relativity theory with grotesquely complicated equations. This does not detract from the validity of Einsteinian physics. The problem of choosing the best or simplest 'frame of reference' is always with scientists of any discipline. The panglossian adaptationists being chided by Gould and Lewontin are just bad practitioners of their science, at the very worst. But their possible excesses should not detract from the validity of the basic tenets of their discipline.

The third minor law of evolution mentioned above was Haeckel's law – that ontogeny recapitulates phylogeny. We have already discussed it in section 10. It has been introduced into modern discussions by paleontologists concerned with the question whether the mechanisms which have been revealed in population genetics and which suffice to explain micro-evolution will suffice to explain macro-evolution as well. Although Haeckel's law has many exceptions, this may be due to a misplaced emphasis in its interpretation. Von Baer has proposed an alternative explanation of recapitulative phenomena. Ontogeny does not recapitulate adult stages of phylogenetic forebears. Rather, the ontogeny of descendants preserves elements of the ontogeny of ancestors. Once more, the general tendency to which Haeckel's law alerts one has become a candidate for some constructive explanation in the synthetic theory. Riedl's systems-analytical approach to macro-evolutionary phenomena (1977) is an attempt to adumbrate the cybernetic design of a genetic programme. He collects a great many broad features of ontogeny and phylogeny that are familiar to taxonomists and developmental biologists, such as the establishment of standardised sub-structures within individuals, symmetries, the co-adaptation of parts that are useless on their own, spontaneous atavism and many others; and reckons to this cluster of questions calling for simultaneous explanation in an all-embracing model of the genetic programme governing individual growth, the Haeckelian question 'What forces the epigenetic system to preserve its own history?'

Thus even if laws such as Haeckel's have not stood the test of time very well as sweeping generalisations, they can still be recognised as containing important grains of truth that should not be lost in the final explanatory story. No doubt their early formulation was too naively general; it would have been far more appropriate to couch them as statistical generalisations, given that they deal with such complex subject matter as the growth and organisation of biological systems.

There are, however, laws of evolutionary biology that have not been framed as inductive generalisations or statistical claims in the light of examined cases. They are laws that can be decisively tested, but which were deduced logically from genetic theory; and, although they deal in probabilities themselves, they have not been formulated on an inductive basis. An excellent example is Fisher's prediction that a stable 1:1 ratio of males to females will be found in sexually reproducing diploid species (1930). Of course, this claim is probabilistic in nature: we cannot with justice take Fisher to be saying that there will always be exactly equal numbers in each sex. The claim was not based inductively on observation of sex ratios in certain species. Rather, it was derived from the synthetic theory as follows. Suppose we have an uneven ratio (say 4:1) of males to females. Then females will be in great demand. Individually they will have better reproductive prospects than males. Thus parents with a genetic tendency to produce females will be selected for, and those genes will increase in frequency, thereby changing the sex ratio in favour of females. This process is obviously symmetrical, and we can expect damped oscillations to converge on an even distribution of males and females. This reasoning is, moreover, quite independent of the creatures' being polygamous or polyandrous or promiscuous or being organised in dominance hierarchies. Fisher's prediction has indeed been put to elegant experimental test (Cherfas, 1981).

An equally striking testable conjecture about sex ratios has come more recently from the sociobiologists Trivers and Hare (1976). Haplodiploid species such as bees and wasps have reproductive arrangements that are not quite so straightforward as those of diploid species. Their males are haploid – that is, they have single chromosomes rather than pairs of them, which are all derived from their mother. She bears them parthenogenetically, that is, from unfertilised eggs. The worker females are sterile. By considering both the pattern of family relations and parental investment in such species, Trivers and Hare were able to predict sex ratios in close agreement with those observed in many different species of social hymenoptera, like ants, bees and wasps. Furthermore, known deviations in vertebrate sex ratios could be explained by Fisher's theory in conjunction with Trivers's theory of parental investment (Trivers and Willard, 1973).

Thus far the burden of discussion has been on Popper's allegation of unfalsifiability or untestability. We hope this has by now been shown to be unfounded. His other main criticism, mentioned above, is that evolution is a singular, historically unique process and as such is not appropriate subject matter for a science whose business it is to make predictions about the future course of that process (Popper, 1957). Halstead (1980) has entered an interesting conjecture as to why Popper should have raised this objection to evolutionary theory. It is of a piece with his rejection of the possibility of historical laws, the denial of which was part of his case against dialectical materialism. Popper was led by his 'life-long revulsion of feeling' over deaths he had witnessed at a demonstration instigated by communists, to 'throw out the evolutionary baby with the murky revolutionary bath water'. But Popper's historicity charge against evolutionary theory is absurd. Pressed to its limit, it would analogously deprive physics of its scientific status. For even though we experiment, at any given time all that has taken place up till then is merely part of a unique, singular, historical process – namely the history of the cosmos. Obviously what the theorist investigates and formulates laws about are regularities within this unique process, laws which support counterfactual conjectures as well as experimentally testable claims. The theory of evolution is just such a theory, as is especially revealed by the possibility of laboratory experiments with large populations of short-lived organisms. To be sure, there are problems of extrapolation to macro-evolutionary process that lie inaccessibly in the past. So there are also in extrapolations of grand unified theories of relativity to the birth of the cosmos. Modern-day cosmologists have problems with their own 'fossil record' of background radiation and the distribution of matter in the universe. Experiments in nuclear physics using high energy accelerators that reveal what particulate interactions characteristically take place, can be likened to the population geneticist's micro-evolutionary investigations in the laboratory. The cosmologist uses his experimental 'micro' findings in astronomy (his 'fossil record') to support various theories about the origin and evolution of the universe. The evolutionary theorist likewise uses his experimental 'micro' findings and his observational findings in paleontology, comparative anatomy, etc., to support his theory of the origin and evolution of species.

Both the basic tenets as well as various probabilistic predictions of evolutionary theory are eminently testable. What it predicts or postdicts usually has to do with patterns or statistical distributions (such as sex ratios); very little is ever said about particular events or particular species in advance of their happening or coming into being. It is here that one needs to emphasise the character of evolutionary theory as that of inference to the best explanation. Events in the past can be assigned explanatory loci without our requiring that one should, before they occurred, have been able to predict their occurrences in every detail in advance. *That* extreme demand cannot be satisfied in physics either. A good example that Stegmüller (1969) adduces in this case is that of a seven-fold collision of billiard balls. Unpredictable quantum effects ensure that the trajectory of the seventh ball struck is wholly unpredictable. Nevertheless, predictions of a less specific kind can be made. The evolutionist can predict with caution that species that are relatively independent of environmental changes, in so far as their members can learn from their experience and modify their behaviour accordingly, will have less chance of becoming extinct than will others that depend very heavily on a special feature of the environment. In Mauritius a species of palm is presently threatened with extinction, since its seeds would germinate only after passing through the stomachs of dodos (which, as is well known, are now extinct). But then, of course, Homo sapiens is the most versatile creature of all, and may well be on the verge of self-extinction. Hence the reason for caution.

Imagining ourselves placed on a twin earth on which the first hypercycle had found itself a protective membrane, all that we could predict with certainty would be that *some* evolutionary changes would take place, and *some* new species would in due course arise. We should not, however, be able to predict *of any* evolutionary change that it *would* certainly take place, or *of any* new species that it *would* certainly arise. The interplay of chance and necessity is revealed in the relative scopes of a modal operator and existential quantifier.

II

CULTURAL EVOLUTION, SOCIOBIOLOGY AND ETHICS

1 CULTURAL EVOLUTION

Culture is the universal condition, if not to say predicament, of man. But it may not be exclusively his. We may distinguish two levels of culture: the first, material, level is that of chattels, artifacts, tools, toys, musical instruments, vehicles, shelters, buildings, roads, dams, monuments, etc; the second, symbolic level embraces rituals, customs, skills, folklore, proverbs, rules of thumb, poetry, music, dance, worship, law, literature, science, and definitions of culture itself. The vernacular sense of culture is usually pigheaded or snobbish, chauvinistic or self-centred in emphasising the tastes, accomplishments, life-style and possessions of the definer. We, however, incline to the anthropologists' more catholic notion, regarding as cultured not only cave dwelling head hunters but also perhaps bands of Japanese macaques frolicking on the beach and teaching their young how to float grains of cereal from sand.

It is almost impossible to say, conceptually, where culture begins. Is all learned behaviour evidence of culture? Some bird songs are learned, and innovations in them are passed from generation to generation by imitation. Are artifacts evidence of culture? Do we count as artifacts such things as birds' nests, beavers' dams and ant hills? Can one discern the glimmerings of cultural practice in the chimpanzee's use of suitably fashioned grass stems to extract termites from their tunnels? In response to such questions, more puritanical definers might wish to stress culture's necessary connection with language. But then the problem shifts to a definition of language, which we shall defer to a later chapter.

The existence of the second cultural level depends on that of the first and on the existence of the organisms that have produced the first level. There are no scientific theories without brains and books; there is no music without mouths and instruments. But one of the oldest controversies in the study of culture is whether symbolic cultural facts somehow transcend the organisms and the artifacts, acquiring casual credentials peculiarly their own.

This is, of course, only to put the matter very vaguely; and indeed much of the problem for the anti-reductionist or anyone supporting the autonomy of higher levels such as the cultural is to specify just what the impossibility of reduction or the autonomy of the higher level consists in. We wish to enter at least a few tentative distinctions that may help to guide subsequent discussion.

First, as already indicated, there is a distinction between material or physical objects, such as organisms and artifacts, and the others supposedly enjoying an autonomous existence, even though those others may in some sense depend ontologically on the former. The most important problem in characterising this second level is: does it consist of entities (or objects) that are clearly identifiable, or is it made up rather of properties, qualities and aspects of material objects, or both? Think of the sorts of 'things' to be subsumed into the second, non-material level: mental states, thoughts, ideas, theories, meanings, beliefs, desires, pains; majority opinions, taboos, prejudices, religious convictions, customs; fraternities, families, firms, organisations, sects, societies, cults, assemblies, cabinets, councils, courts and markets. Quite obviously there is a plethora of second-level 'things' of sounder or more dubious individuality.

As a first step towards bringing order into this confusion, let us briefly if dogmatically note the following. Organisms like human beings can behave in quite complicated ways, interacting with their environment and with their fellows. That complex behaviour is the ground for ascribing to them certain 'mental states' and dispositions – characterising them as having certain beliefs and desires and preferences, and understanding certain conventional or 'linguistic' forms of behaviour in certain ways. We also ascribe to these beings an 'inner' or 'mental' life consisting of various episodes and processes – sensations of hunger, pain, satiation, pleasure, and periods of contemplation, attention, intellectual effort, and so on. All this is familiar and rich farmland of the philosophy of mind. Mental states, dispositions, processes and events are, in the broadest sense, properties of a behaving organism; they are not strictly speaking *things* that have a life of their own, to which we can refer and which we can confidently re-identify, as we can the creatures that 'possess' them. Philosophers today are loath to reify meanings: sentences or utterances may be *meaningful* (by virtue of their place within a pattern of use) and may indeed be meaningful in a very rule-governed and orderly way that even affords the prospect of a scientific semantics and speech-act theory, but that is not to say that there are things called *meanings* that sentences or utterances possess or enjoy. Likewise the attribution to someone of a certain belief that such-and-such is the case (a belief expressed in some familiar language) does not justify the view that beliefs are things that people have or carry in their heads; such attributions only provide aspectual glimpses on the person's behaviour or dispositions to behaviour. They are part of an explanatory and predictive apparatus that helps us to make sense of, to understand, what the person says and does. And as Davidson (1973) has made philosophers so crucially aware, attributions of belief, desires and grasp of meanings are constrained by holistic considerations; the three components of the explanatory scheme must be mutually adjusted in the light of observed behaviour. Thus in entering the realm of the mental – which is the same as to say the realm of the intentional, and of the intensional or symbolic – what we encounter is not a new level of non-material entities or things, but rather a new level of *description* of what first level things are getting up to. That is a first step in clearing up potential confusion.

The second step is to note that among the second level 'things' we kept certain ones separate towards the end – fraternities, families, etc. These are what may broadly be termed 'social individuals'. They are individuated not just as sets or collections of their individual members, but as functioning aggregates that derive their integrity, coherence and identity from the mutual relations and interactions of their members. There is a possible analogy here with the way individual organisms are composed of living cells – not as mere assemblages of them, but as functioning aggregates whose mysteries biology has long sought to disclose. Now it is the organismic philosophers of biology (such as Uexküll (1928) and Bertalanffy (1972)) who maintain that the complicated biological functioning of an organism is somehow 'more than the sum of the behaviour of its parts' – whatever that is supposed to mean. Likewise in sociology and anthropology there are some thinkers who regard what we have just called 'social individuals' as some sort of *superorganic* entities, entities whose actions and functions cannot even be *causally* understood in terms of the sum, statistical or otherwise, of the behaviour of their individual members. Thus not only do such thinkers (such as Durkheim (1938) and Radcliffe-Brown (1952)) probably insist on the new, second level of *description* (the mental and symbolic); they now also insist on a new, second level of *things*. These superorganic entities are referred to not just by way of convenient definitional or abbreviatory shorthand, not just as collections of their members satisfying certain simply statable conditions; they are rather referred to now, out of theoretical necessity, as things in their own right, things with their own causal powers and roles, powers and roles that cannot be decomposed in any definable way into those of their constituent members. This is holism of a quite different sort from that of Davidson mentioned earlier. For Davidson insists on holism as a constraint on the formulation of explanatory theories at a certain level of description. The constraint calls for simultaneous attribution of the main notions being thus holistically orchestrated in the light of all the evidence. Yet he still insists that all that there is, is material, and all that happens in the material realm admits of first-level, material explanation. Durkheim, by contrast, enjoins a holism that might better be dubbed 'wholism'. For him, there is no complete underlying causal story on the material or physical level; rather, once the level of the 'social fact' comes into existence, it acquires a unique causal potency, essentially supplementary to that of the material level. The irreducibility of laws of social science, for Durkheim, is then not just a Davidsonian matter of their being laws formulated at a different conceptual level, providing a different kind of explanation of phenomena that nevertheless simultaneously admit of exhaustive physical explanation. Rather, they are irreducible because they are laws of social things with social properties – things that are not merely definable collections of individuals, and properties that are not merely the outcome of interactions of those individuals.

We do not see just what philosophical gain there is to be had from a Durkheimian insistence on wholistic social existents in the sense just explained. For one thing, it complicates the picture that must emerge in the light of any evidence that may be discovered to the effect that Durkheim wholes *do* exhibit intimate, law-governed dependencies on their constituents and the latters' behaviour. Should such evidence emerge (and there are reasons already to think it has) then the Durkheimian will have to resort to an interactionist thesis, a thesis about the interaction of social and biological organisms. Indeed,

Durkheimians may already be in this position. But, the more penetration is revealed between individuals and the superorganisms they make up, the greater is the prospect of dissolving the superorganisms definitionally into their constituent individuals and the bonds among the latter. Even if only for methodological reasons, it is profitable to regard the superorganisms and their behaviour as 'the statistical product of the separate behavioural responses of large numbers of human beings who cope as best they can with social existence' (Wilson, 1978).

Note that it is nothing peculiar to *cultural* existence that produces superorganisms with 'wholistic' pretensions. An organismic biologist may pass beyond the level of the individual organism, and extend his doctrine to the case of, say, a termite colony or a slime mould, even though he himself might agree that culture does not yet come into the picture. Likewise, someone may apply Davidsonian ideas elsewhere in the natural order, and be prepared to attribute to languageless creatures certain proto-beliefs, and proto-desires in accordance with holistic constraints, provided only that the creatures concerned behaved in an appropriate way. Davidson himself, however, insists strongly that beliefs, desires, and command of a full-blown language come all in one package, and that attributions to other creatures of any particular belief or desire can only be made when one is able to 'read into' their behaviour a system of linguistic transaction every bit as rich and complex as our own. We shall have more to say about this in a later chapter. In sympathy here with Davidson, one cannot help but notice the extent to which the 'symbolic level' of culture, of which examples were given above, depends on the community of organisms concerned having a language. If one is prepared to regard certain behaviour as constituting a ritual or the enactment of a law, or an initiation, or a formulation of a common plan of action, it is impossible to refrain from the temptation, or escape the obligation, to interpret some (perhaps much) of the behaviour as *linguistic*, as conveying meanings or being meaningful, as being part of a system of communication mastered by the participants concerned. But as we remarked above, we are reluctant to get involved in legislative definition. Culture is incipient in the very crude tool-use of chimpanzees, and the way their young learn that use by example and imitation, even if they do not fulfil the most stringent requirements put up by McGrew and Tutin (1978). Perhaps our willingness to let this in as a case of culture has something to do with the specificity of what is being learned by a highly malleable, plastic and educable young chimpanzee, and also something to do with the fact that their behaviour appears to support attributions of proto-attitudes and motivational states, and even the recognition of those attitudes and states in their fellows.

They can transmit a practice from one generation to the next; and this practice involves complex sensory-motor skills, more complex than that in the case of 'learned' bird songs. (Interestingly in this connection, Mundinger (1980) regards the diverse songs of colonies of eastern house finches as the result of *cultural* speciation.) The motivational states underlying the practice are easy to discern, the obvious goal important enough to explain why the practice is inherited by imitation. A medium of transmission outside the germ line has been created – a necessary, even if not sufficient condition of cultural inheritance.

The existence of suitably sophisticated artifacts likewise invites interpretation as evidence of culture on both the criteria emerging from our discussion. The first criterion is the presence of language, which allows the individual experiences and memories of one organism to be put at the disposal of others, and which allows them to formulate joint purposes and engage in joint action, where these purposes and actions cannot be attributed to closed genetic programmes. The second criterion is a new medium of behavioural inheritance, one which is perhaps *ipso facto* present when the first criterion is met. When archeologists and paleontologists discover remains of tools and shelters bearing the mark of sophisticated plan and execution, they make the reasonable inference that only creatures who could communicate by language could have produced them. Likewise, it is often assumed that only linguistic creatures could have succeeded in the primitive hunting missions for which evidence has been found (cf. Leakey, 1981c). For perhaps these reasons the anthropologist Sahlins (1976, pp. 13-14) is prepared to assert that culture

> is itself considerably older than the human species as we know it. Culture was developed in the hominid line about three million years ago.The modern species of man, *Homo sapiens*, originated and gained ascendancy about one hundred thousand years ago. It is reasonable to suppose that the dispositions we observe in modern man, and notably the capacity – indeed, the necessity – to organise and define these dispositions symbolically, are effects of a prolonged cultural selection.

Our own approach to the problem of culture, its origins and evolution, will be both cautious and tolerant in the way of definition and delineation, but rigorously physicalist in the way of method. Organisms and their physico-chemical constituents, for our present purposes, are the only things that we shall acknowledge as existing. Mental states will be understood in the way described above; social individuals will be regarded as law-governed assemblages of biological individuals. Yet we can still acknowledge the autonomy of the mental and the cultural in a Davidsonian spirit, by seeing mental and social terminology as affording a special level of description and explanation not reducible, as theory, to the physical level of description. That is not yet to say, however, that what goes on is not entirely physical; it is only to say that there is not much hope of understanding it profitably as only physical. Nor is it to say, as Winch (1958) does, that we cannot apply a scientific method in pursuit of higher level laws. But we can go a little further than Davidson, and admit the possibility that there may be law-like regularities (even if only statistical in nature) involving both the physical and the mental-cultural level of description. This indeed must be the case with any theory of 'feedback' or 'interaction' between genes and culture. The autonomy of mental or cultural description does not entail the nomic insultation of the mental and cultural realms; there can be 'bridge laws' usefully relating mental and cultural states with physical substrates. While maintaining that the latter are all that exists, we can nevertheless acknowledge that the insight into 'feedback' and 'interaction' between genes and culture would be wholly lost if culture had to be described in completely physical terms.

Our main concern in the present section is to reach some understanding of cultural evolution, and the extent to which it can be understood along the lines of biological evolution. In the latter case we have clarified the notions of genotype, phenotype and environment and explained the three important conditions of variability, heritability and differential reproduction. Adaptation is regarded as the long drawn outcome of natural selection winnowing out certain phenotypes so as to leave behind such genotypes as will, in the environment(s) concerned, enjoy reproductive success. In the phenotype we have discerned both morphological and behavioural traits; and the latter are adaptive just in case their presence in past ancestors helped to bring about their presence in present descendants.

The great problem with the emergence of the cultural is that it would appear arbitrary not to reckon skills, accomplishments and capacities that we would wish to characterise as cultural to the (behavioural) traits in the phenotype as it is already understood for the purposes of biological evolution. Likewise, culture imbues the environment; and it is almost futile to ignore it, and to look for a 'purely biological' environment, when talking of environments as already understood for the purposes of biological evolution. *A priori* biological evolution is going to produce culturally adapted individuals when an important part of the environment is cultural; and genotypes placed in such environments will produce phenotypes with important cultural traits. It is a dogmatic white lie to maintain that 'the biological factor of man is irrelevant to various problems of cultural interpretation such as diversities among cultures, and processes of culture change in general and the evolution of culture in particular' (L. A. White, 1959, p. 12). Genes, however, even when arranged in new genotypes better adapted to a world of culture, remain the boorish self-replicators they have always been. Nevertheless, to discern a causal interaction between genes and culture is not to make a category mistake; indeed, its denial is a logical contradiction once culture is reckoned to the environment, cultural traits are reckoned to the phenotype, and the core conditional of evolutionary theory is understood just as before.

This is all there is to cultural evolution from the austerely physical point of view. Past phenotypes behaved in certain ways, ways that grew more complex through the development of open programmes. The complex behaviour resulted in certain conventions, which conferred upon behavioural patterns certain meanings. Some of the behaviour left concrete remains in the form of artefacts. Behaviour brought these into use. Artifacts, their methods of use, and patterns of conventional, linguistic behaviour came to form an ever-increasing part of the environment of new genotypes. Their mastery of tools and meanings was of great consequence for reproductive success. Genotypes better adapted to that mastery consequently evolved. Tools begat tools in the hands of their makers. The face of the earth began to take the shape of a cultural portrait. Archeology scratches its surface, seeking the earliest pigments, the fossil record of culture itself.

'Part of the environment, part of the phene' – thus should culture consistently be understood. And this account of 'cultural evolution', which in our opinion remains the best, really countenances biological evolution as the only evolutionary process properly so-called, with the environment becoming ever more cultural, phenotypes becoming ever more acculturated, and gene frequencies changing accordingly. There is positive feedback between genes and culture without having to reify culture, either super-organically or through its linguistic and symbolic constructs. We must constantly bear in mind that it is the biological level that subtends the cultural, and not the other way around. Any 'law' of cultural evolution that predicted ever-increasing complexity of artifacts or linguistic constructs or social groupings would depend for its truth on a law of biological evolution guaranteeing the genotypes that could live in and sustain the more complex cultural environment thus predicted. If all the members of a community die, then so does their culture; but if their culture dies out, then at worst they return to a state of nature.

Some thinkers wish to accord culture a life of its own, subject to its own evolutionary process parallel to (perhaps interacting with) the biological one. They would dignify culture as more than just aspects of human bodies and their surroundings. One might be inclined to this view of matters after making initial observations such as the one we made above, to the effect that culture leaves its own fossil record of artifactual deposits. One may be impressed with the apparent 'phylogeny' of a product such as the motor car, such as is revealed in museum displays of various models in the chronological order of their invention and production. It has been remarked that the earliest motor cars were shaped very much like horse carriages, although, of course, there was no inherent design reason why this should have been so! (See other examples in Eibl-Eibesfeldt, 1973). One may be tempted to see a clear parallel with, say, morphological changes through some phylogenetic line of mammal, like the well-worn example of the horse. But the question is whether this in itself would elevate the word 'evolution' in the case of artifacts to its more powerful role in the realm of biological explanation.

Some 'wholists' and autonomists wish to go further. For them not only artifacts, the material products of culture, have their archeology and implicit phylogeny. So also do the more ethereal products of culture, the denizens of Popper's 'third world' (1974) – ideas, insights, creeds, inventions, theories. These may come to be fossilised in the material products of the cultures concerned – on clay tablets, tomb walls, scrolls, books, records and magnetic tapes. They are first and foremost linguistic constructs, and their physical fossils require an act of linguistic interpretation to recover them. Not only do material products and linguistic constructs of culture 'evolve'; so do also institutions, social structures, skills, modes of production, techniques of war, legal procedures, methods of management and so on. That these have all emerged from a process of variation and 'selection' is one of the reasons why Hayek (1978), Lorenz (1970) and Campbell (1975) enter what may sound like a conservative plea for tolerance of the institutions we have. For they may, in ways perhaps difficult to understand or 'read off', embody an evolutionary wisdom exceeding that attainable by a wise man in a normal lifetime. Another extensive application of the evolutionary way of thinking is evident in current literature in theoretical economics on the business firm. (An able survey of the field is to be found in Hirshleifer (1977).) Ghiselin (1974), in turn, enriches his individual selectionist theory of the 'sexual economy' with self-consciously sustained analogies with business economics. It is clear that an extensive analogy may be drawn between economic systems and biological systems, with economic analogies of such notions as reproductive survival, adaptation, exploitation of resources, predation, reproduction, mutation and diversification readily forthcoming. Malthus (1798) got his metaphor from biology: Darwin borrowed it back; social Darwinists crudely retrieved it; and theoretical economics is now rehabilitating it in a descriptive rather than normative mode.

If one shifts from the process of biological evolution in its cultural phases, in which individuals and their environments bear the marks of culture, to a parallel process at a distinct, 'cultural' level, in which cultural entities themselves undergo evolutionary change, then the burden of analogy is a heavy one. Let us explore how profitable it is to postulate a separate cultural evolutionary process moulding cultural products to produce further ones in ever more 'adapted' forms.

The first problem is to identify what the cultural entities in question are, and to explain the sense in which they 'reproduce' themselves. One seeks a forebear-descendant relation of some sort, even if not a parent-offspring relation. One searches for pedigrees, for lines of descent, for some mechanism of inheritance. Now the forebear-descendant relation and the associated mechanism of inheritance may be different, depending on one's choice of cultural entity. Literary theorists can seek the pedigree of a poetic motif, seeing the mechanism of inheritance as one of sensitive and self-conscious 'quotation' or allusion to famous sources, one that can be made and understood only by those steeped in a literary tradition and sharing a vast background of tacit assumptions. Social theorists can seek the pedigree of institutional structures or political concepts; and here the mechanism of inheritance may be much more 'collectivist', for institutions can outlive the people in their corridors, and can be created or changed by committee rather than

Napoleons. Historians of ideas may seek the pedigree of a political concept. An influential tract may have been influenced seminally by earlier sources, quoted or acknowledged or not. Finally, industrial archeologists may seek the pedigree of something like the smelting pot, investigating its spread through communities that were in contact with one another, or exchanged entrepreneurial artisans, or lost them as refugees, and so on.

Influential ideas, material artefacts, social structures: all three kinds of cultural entity reproduce themselves in subtly different ways and have their own peculiar mechanisms of inheritance. All these types of inheritance, however, are both lateral and vertical; whereas biological inheritance is only vertical. Even if one were to concentrate on artifacts alone there would be ambivalence over the best analogy with biological evolution. One may wish to regard the actual physical products as the entities undergoing evolutionary change; or instead take the 'design idea' underlying them as so doing. In the first sense, wax moulds of existing artifacts such as knives, used to cast more knives of the same (but now perhaps slightly different kind) would represent a pretty immediate sense of 'inheritance'; but this is a special case. The seeds of inspiration and the fruits of invention are often much further removed. So-called 'second generation' tools and technology are related to their first generation forebears not by means of individual 'begettings' as in the case of the wax mould knife, but rather by way of wholesale innovative shifts in design type.

Let us now attempt to press as thoroughgoing an analogy as we can. It is not enough to speak of cultural 'evolution' if all we have in mind is gradual change of artifactual or institutional design, or of the cast of an idea. We must rather seek a 'pool' of entities on the cultural plane that will correspond to the genepool in the biological case. We must make sense, as mentioned above, of their replication and mechanism of inheritance. We must further account for mutations and their selection, leading to adaptation. And finally, it would be useful also to be able to discern a contrast between the genotype and the phenotype of a cultural entity.

In the biological case the gene, although physically embodied as DNA, is nevertheless important in its *informational* role. The information it carries has to be realised, expressed, extracted during ontogenesis in the synthesis, first and foremost, of functional proteins. Alleles compete with one another for representation at a locus. We shall now exploit an idea popularised by Dawkins (1976), but anticipated by

Burhoe's notion of a culturetype (1967). Dawkins suggested, as the cultural analogue of a gene, the information notion of a *meme*. We shall go further, and introduce also *memeles* in analogy to alleles! To give examples immediately, for the meme 'calculating device' we would have the memeles *slide rule*, *abacus*, *hand held electronic calculator* and so on. For the meme 'wheel' we would have *solid wooden wheel*, *spoked wheel*, *iron rimmed wooden wheel*, *rubber lined metal wheel*, *wheel with pneumatic tyre* and so on. Now, of course, it is difficult in certain cases to say exactly where one meme ends and another begins when componentry comes into the picture. Why the memeles just given in the case of 'wheel', rather than more specialised ones for 'spoke', 'rim', 'axle joint', etc.? Our answer here is that the gene concept as needed for explanations by natural selection is certainly not as sharp-edged as that provided by molecular genetics. One speaks of genes 'for' altruism, when indeed there must be a great many genes in the sense of molecular genetics involved in the development of altruistic traits. Likewise, the genes 'for' almost any adaptive morphological trait are in the strict sense polygenes, but are treated as units in explanations by natural selection. So we might tolerate a certain fuzziness in the meme concept as a necessary hazard in giving a particulate model of cultural phenomena. This is not to say that we are particularly enamoured of a particulate model; but any model of cultural inheritance will have to see molecules in the winds of change. How are memes – the informational units – physically embodied? First and foremost, in neuronal memory circuits in living brains. One and the same brain may contain several memeles for one locus, by virtue of acquaintance with the ideas concerned. The memele that 'expresses' itself phenotypically, that is, the dominant one, is the one advocated or applied (in the case of artifact design), or the one asserted or proved (in the case of theoretical science), the one chosen for performance (in the case of poetic or musical memes). The cultural phenotype of a person is thus a shifting, evanescent collage of presently-dominant memeles. Moral ideals, theoretical convictions, everyday beliefs, skills, accomplishments, tastes and inclinations in food, dress, music and literature, political attitudes and a whole lot more all go to make up the cultural phenotype of a person at any given moment. Some aspects of this phenotype will be relatively stable, others more volatile and labile. He may be easily led to change his views on a topic, to adopt a new kind of tool, to acquire a new method for solving certain problems, to re-order his preferences for certain consumer products. The cultural phenotype is thus embedded within the biological phenotype construed in its broadest sense. Memes recessive now may be dominant later, in one and the same lifetime. Brainwashing and propaganda are forms of meme engineering. The way memes are transmitted and replicated is by communication with other brains, inculcation, teaching, example and imitation, oratory, scientific congresses, letters to the editor and so on. Memele frequencies change in the meme pool as a result of rapid lateral transmission as well as vertical transmission to one's children – who, in times of accelerated cultural evolution may all too often grow up with very different cultural phenotypes to those of their parents! One might try to exterminate a meme by killing off the bodies that house them; but ideas have a way of outliving their bearers, especially when they can lie dormant in an archive, awaiting rediscovery, decoding, and further transmission. The rate of meme mutation can increase when conditions of stress require it to – as, for example, at times of war, when the military effort can throw up many new and useful technological memes; similarly for the cancer research programme and the space programme. But this is still fundamentally like the biological case, where there can be selection for increased mutation rates. Mutations of memes might very well be 'random' with respect to selective needs, despite appearances to the contrary. Many an idea is consigned to the waste paper basket as scientists and engineers collaborate in the solution of a pressing problem. At present too little is known about creative intellectual processes to tell whether there is a subconscious random idea generator whose products are screened and selected still at a subconscious level before entering the conscious mind. If this is indeed the case, then we would have an impression of 'non-random' mutation which was nevertheless mistaken. There is some evidence for this picture of intellectual invention; take, for example, Kekulé's discovery of the structure of the benzene molecule after dreaming of a snake swallowing its tail, or Poincaré's discovery of the series of Fuchsian functions, which came to him in a flash as he boarded a charabanc for a Sunday outing after drinking thrice boiled coffee.

What other elements of analogy with the biological case can we trace? Pipes, wheels, cogs and bricks may be regarded as pleiotropic memes; and no doubt one can see polymemes in componentry. The disanalogies are the following:

1 Lateral transmission to members of the same generation (as already mentioned above) resulting in much more rapid change than in the

biological cause.

2 The number of memes, and of memeles for a meme locus, can vary from person to person. A polymath carries a large meme complement with a large degree of polymorphism; a dullard only a few.

3 The memotype can change within a person's lifetime, owing to forgetting and new exposure.

4 The material embodiment of a meme, and detection of recessives, may be a more complicated matter than in the case of genes.

5 There is a *quantitative* difference in mutation rates of memes and genes, but meme mutation may still be a random matter (see above).

6 Perhaps meme recombination is Lamarkian (i.e. all known memes can be recombined at will and as the need arises) even if meme mutation is not, although the distinction between mutation and recombination may at times be unclear.

7 Highly maladaptive memes, when dominant in a person's cultural phenotype, can be eliminated from him without causing his death – by therapy, persuasion, indoctrination.

8 There is nothing in the cultural case analogous to chromosomes; and whole memotypes reproduce only in a very tenuous sense (but may happen in the case, say, of a primitive tribesman taking wholly after his father).

Just as genes must fit into a whole background of genes, and the fitness of alleles is to some extent dependent on the rest of the genotype in which they find themselves, so also with memes. A new idea, a new hypothesis, a new proposal for political reform, a new technological invention, is tested by selective forces in part due to the already existing background of memes in which it arises. To a large extent, especially with memes of scientific theory, the forces of selection are those of the critical, rational, experimental method. Especially when memes are likely to contribute to the biological fitness of their bearers, there will be intense selection for true and simple ones – as witness the growth of medical science, and the spread of every kind of technology that increases longevity. Other memes may be 'easy riders' – such as literary figures deriving from the governing technology of the day, and political and aesthetic memes merely 'reflecting' the memes involved in industrial modes of production and their control.

The selective filter through which memes must pass has both a rational and an emotional component. And to a certain extent these interpenetrate: as Kuhn (1970) pointed out, scientists can become so emotionally committed to certain theories that they cannot abandon them when they are falsified; and almost every reader will have experienced the intense emotional and aesthetic pleasure that can be derived from artifacts that serve their functions perfectly. As Pulliam and Dunford (1980) point out, an important selective agent for memes must be the phenomenon of 'cognitive dissonance'. People constantly check different parts of their world-view for consistency. New ideas are much more acceptable on first acquaintance when consistent with beliefs already held. The ultimate selective test for memes, however, must surely be biological survival – survival of enough living brains to carry them into posterity. A highly maladaptive meme once took hold in a Russian cult – they subscribed to a ban on sexual intercourse. Nor did they make any converts. They died out, as can be expected, and so did the expression of their meme with them. Likewise, the early kibbutzim ideological meme of no gender differences in political and domestic matters succumbed to a recessive status as genetically based dispositions to sexually differentiated behaviour reasserted themselves (Spiro, 1979). Feyerabend's plea (1975) for what amounts to high mutation rate for scientific memes may be well and good, but a mutator meme in the moral sphere, an anarchy meme of kinds, may be deleterious. Campbell (1975) has argued that here, as in biological evolution, there must be a balance between frequency of mutation and faithfulness of replication, on pain of break-up and disintegration of the social system whose very backbone is coded for by moral memes. Moral fibre does not consist of proteins alone.

The early kibbutz movement subscribed to an ideal of sexual equality which was to be achieved through sexual identity. There were to be no gender differences in the rearing of children, domestic chores, farm labour and other economic occupations. They established collective nurseries where parents would have equal, if short, time with the children. Despite the fact that highest status attached to farm labour, what transpired was that the kibbutzim women of their own accord reverted to their more traditional roles. They demanded more time to mother their children and less time toiling in the fields. They grew more disinclined to involve themselves in the political affairs of the kibbutz. Spiro has sympathetically and dispassionately studied this social phenomenon, which he describes as 'the [women's] counter-revolution in sex-role differentiation'. He concludes that 'the counter-revolution initiated by the [women] in the sex role system, like that in the family system, was determined to a large extent by sex differences in precultural motivational dispositions'.

The case of the kibbutzim raises a theme of great importance: how far does man's basic biology constrain possibilities of cultural innovation and change? One might be prepared to concede that the women rebelled in the kibbutzim because of their earlier steeping in wider Jewish culture, with its emphasis on the family and distinct maternal and paternal roles. But many of the 'rebels' grew up and were socialised in the kibbutz culture, though, of course, they would not have been wholly innocent of the wider Jewish culture outside. Moreover, one can press the question one step further back: how did that wider culture arise, and what explains its stability? Even if the pattern of family relations was imposed by men upon women at its inception (which, of course, is highly unlikely), then we would still have to face the possibility, if not probability, that there has since been genetic selection for those personalities best adapted to those cultural arrangements.

Changing cultural conditions constitute selective forces that can radically alter the genetic complexion of the people concerned. Given present heritabilities (in the second sense) of many important cognitive and personality traits there is no reason to regard them as non-heritable in the past, or they simply could not have evolved. We must consequently acknowledge the likelihood of sustained micro-evolution; and as we saw in chapter 1 this can effect considerable changes within relatively few generations. The greater domesticity of women in all cultures, despite intra- and inter-cultural differences in this respect, might, implausibly, have been imposed upon them by superior male force and might in no way have been tied to their genetic constitution. But a few millennia of this selective regime would have brought forth daughters who were reproductively more successful, because more happily adapted to their domesticity.

Suppose a community of cloned (that is genetically identical) individuals were established somewhere, with no genetic difference between males and females apart from direct sex-determining genes. Any son would be genetically identical to his father, any daughter to her mother. They are visited by a Martian who imparts certain knowledge and inventions. One can assume that cultural changes would result. There would be cultural evolution as these inventions were applied and modified, even though no genetic variation were available. Now to be sure this cultural evolution should proceed more slowly than is normally the case in a population with genetic variability. They will not have, presupposing equal population size, their Einsteins, their Darwins, their Marxs or any innovators who owe their talent to a rare combination of genes. Thus cultural evolution is theoretically possible without a parallel biological evolution. Normally, however, the two evolutionary processes are intimately bound up with each other. Some genotypes make better cultural innovators than others, their innovations change the environment, and this in turn exerts new selective pressures on genotypes. The invention of spectacles, for example, much allayed selection for keen sight, and indeed primitive peoples generally have a lower incidence of visual defects (Post, 1962).

Some memes are bound to be recessive for genetic reasons. The population control meme is a case in point. Thus Hardin (1968):

> It is a mistake to think that we can control the breeding of mankind in the long run by an appeal to conscience . . . People vary. Confronted with appeals to limit breeding, some people will undoubtedly respond to the plea more than others. Those who have more children will produce a larger fraction of the next generation than those with more susceptible consciences. The difference will be accentuated, generation by generation . . . the argument assumes that conscience or the desire for children (no matter which) is hereditary – but hereditary only in the most general formal sense. The result will be the same whether the attitude is transmitted through *germ cells*, or exosomatically.

The plant geneticist Darlington sought in a monumental study (1969) of major civilisations to show that genetic adaptation was the key to cultural success and that successful cultures owed their success in part to hybrid vigour resulting from intermarriage with subjugated peoples or neighbours. In debates with colleagues, particularly Julian Huxley, he accused them of reluctance to pursue the conclusions to be drawn from their own theoretical principles: ideological prejudice led them to insist on cultural evolution transcending and outstripping biological evolution. Perhaps Darlington's meme was just out of joint with the memome (*Zeitgeist*) of his time.

We have given above examples of memes that can be suppressed or restrained by genes. Biological reproductive success is the ultimate arbiter for memes as well as for genes. There are, however, intermediate levels at which memes are filtered out short of that final sanction. One is that memes can cause their potential hosts feelings of satisfaction or dissatisfaction, the biological mechanisms underlying which could have been programmed by natural selection in our hominid ancestors. For example, well-designed axe heads and bows were of crucial survival value to our hunter gatherer forebears; those who 'had an eye' for the well-turned product might, through their consequent selective advantage, have passed on to their descendants a predilection for those simple aesthetic pleasures to be had from taut curves and solid contours. The generalised aesthetic preferences in design that we are here suggesting may have been selected for could depend on the perceived match between any artifact and the purpose for which it is fashioned. Perhaps it was for this very reason that Darwin himself spoke of animals' 'perfection of structure and coadaptation which justly excites our admiration'.

Complementary considerations are advanced by Ziff (1981, pp. 519-20):

> It is essential to realise that genetically based capacities each of which has an explicit utilitarian value may combine to yield a non-utilitarian value. First self-awareness has a basic role to play in the development of a recursive technology . . . [and] is implicated in the recursive structuring of linguistic systems. Secondly the capacity to respond to physiognomic characters may be of utility . . . [as well as] a capacity to appreciate the structure of a complex entity. . . . If in some pause during the struggle for existence each of these capacities is exercised simultaneously under the impetus of an encounter with a provocative entity the result may be the performance of an act having a non-utilitarian value, an intrinsic value, an aesthetic value. Aesthetic value then is a pot of honey found along a thorny biological path while the wolves are at bay!

On the rational side, memes will be tested by intellectual mechanisms for their fruitful coherence with other belief memes, and their logical consequences tested. The logical processes that accomplish this are themselves part of a biologically evolved repertoire of the human cognitive apparatus – about which more is to be said in the next chapter.

Having done our best to pursue the analogy between cultural and natural selection as far as we can, we nevertheless wish to take stock by asking why it is necessary to consider a separate evolutionary process for culture at all. Can one not simply say that the so-called memes are just parts of the environment or states of biological organisms undergoing only biological selection? From our consistently physicalist standpoint we have suggested that memes may be nothing more than functional configurations of neuronal connections. Thus each gene simply finds itself in a much more complicated environment than the ones prevailing at the dawn of culture – an environment, however, which is, from that physicalist standpoint, a mere product of physical processes. Likewise, for those who subscribe to causal theories of perception, belief and action, the memes that reign supreme in any epoch inhabit the heads they do as a result of purely physical processes.

2 SOCIOBIOLOGY

2.1 Inclusive Fitness and Kin Selection

Evolutionary genetics has always recognised the 'evolutionary sense' of parents' caring for offspring, even at cost to themselves, as the genes' way of ensuring future propagation. Sociobiology further recognises that genetically engineered concern will probably extend to other relatives besides immediate offspring. Not only by caring for one's own offspring can one propagate one's genes; but also by caring for other relatives. The appropriate measure of care can be calculated from the degree of relatedness between the caring and the cared for. (We hasten to add that 'appropriate' is meant here in the sense afforded by evolutionary explanations of gene frequencies, and not in any moral sense.)

The degree of relatedness between full siblings is 1/2. If two of my brothers are certain to die unless I go to their rescue, and if that act will save them without making my own death certain, then this latter consideration will in all likelihood tip the balance and make me act accordingly. For doing so ensures maximal representation of my gene complement in the future.

The fact that an individual's genes are also to be found in varying proportions in relatives or kin has made it possible to formulate genetic calculi of social living. Sociobiology considers the advantages, from an evolutionary perspective, that accrue to an individual's genes through forms of behaviour and social organisation. The sociobiologist is interested in mechanisms that maximise 'inclusive fitness' – that is, the likelihood of the propagation of the individual's genetic material, not just by his own offspring. Since this genetic material is also in his kin there is a subtle shift of emphasis from past egocentric conceptions of the struggle for individual survival towards a new notion of kin selection.

In the kin selection paradigm it is the genes of the inclusively fittest individuals that will propagate.

Before matters were seen in this light, biologists had difficulty explaining so-called altruistic behaviour among animals and man. Hamilton's work (1964) made it clear that an individual should help its kin, because it is, in Dawkins's words (1976) a 'survival machine' for its genes. These genes, simply because they have survived so long in the face of aeons of selective pressure, must be the sort of genes which instruct their survival machines to behave in such a way as to maximise the chances of those genes' future replication.

In this new calculus of behavioural strategy and social structure, an important notion is now that of degrees of relatedness. As remarked above, full siblings are related by 1/2. How is this to be understood? It is not a measure of the proportion of loci throughout the chromosomal complement at which they have the same alleles. For at some loci, namely those invariantly lethal upon mutation, the species might not be polymorphic. That is, some loci might have only one allele. Indeed, structural proteins, for example, show little genetic variability, as disclosed by experimental biochemistry. What makes individuals unique is not so much their particular complements of alleles well-represented in the gene pool, for these alleles are carried by many of their fellows. It is rather a handful of relatively infrequent alleles in each individual that makes it unique.

Two parents generate family resemblances through the various combinations of these alleles in their offspring. Full siblings are related by 1/2 because that is the *conditional* probability, concerning any allele, that, *given* that it is in one sibling, it will be in the other. Every paternal and maternal germ cell results from 'random' choices of one allele from each pair at a chromosomal locus. So any allele in one offspring will with probability 1/2 be carried by his sibling.

King and Wilson (1975) discovered an astonishing similarity between the gene pools of chimpanzees and men. The average human protein is more than 99 per cent identical to its chimpanzee counterpart. Since then many critics of sociobiology have claimed that kinship calculations are all irrelevant because human beings are so similar to each other that the small additional similarity owing to close kinship could not give rise to any selective effects. As seen in I.6, the polymorphisms investigated by population geneticists have for methodological reasons frequencies of 5 per cent and more. The fact that the mutation frequency is of the same or a similar order of magnitude as the number of genes an individual contains ensures that most individuals will carry a small number of alleles that are unique to themselves or almost so, depending on population size. These alleles have to be ignored by biochemists investigating genetic variability because they are too rare and their study would involve unmanageable sample sizes. Some such alleles the individual will have inherited from his parents rather than acquired as a new mutation of his own. Some of these alleles (and perhaps some such mutations) he will pass on to his offspring, thereby ensuring family resemblances. Most mutations, we know, are pleiotropic and affect many morphological and behavioural characters. Most are detrimental to their carriers. This explains their low frequency and the considerable variance between families in such fitness characters as longevity. There is no better predictor for a person's life expectancy than the age at death of his parents and grandparents (Scheinfeld, 1965). The proportion of favourable mutations among all mutations is low. Natural selection quickly pushes a newly arisen favourable mutation through the whole population. Thus particularly healthy individuals are best characterised not so much by the presence of favourable mutations but by the absence of unfavourable ones.

The probability that any allele found in individual A will also be found at the same loci in individual B is the degree of relatedness between A and B. Between parent and child this is 1/2, between grandparent and grandchild 1/4, between uncle and child 1/4, between first cousins 1/8 and so on.

How is the degree of relatedness involved in evolutionary explanations? Let us suppose that a new allele arises. It has the greatest chance of being replicated in the future if it helps to construct bodies that ensure its own replication. In particular, such bodies should help other bodies likely to contain copies of that gene to enjoy reproductive success too. But the effort involved in helping those other bodies to enjoy their own reproductive success should not be spread around indiscriminately. The best bet is to help a full sibling; then a nephew or niece; then a first cousin, and so on. Ego's sacrifices of individual advantage are best invested with an eye to the beneficiaries' degree of relatedness to him; or so it is eminently reasonable to suppose from the point of view of the rare gene whose reproductive survival is at stake. Not only would that gene have to programme bodies to lend a hand to others with an eye to their degree of relatedness; but also, with an allowance for their own prospects, given their present age, health, etc., of having further descendants or, in turn, helping their own kind to have further descendants, and so on.

The upshot of this reasoning is that a rare gene would be particularly successful at getting copies of itself into subsequent generations of the genepool if it turned its bearer into a 'kin-directed altruist' behaving *as though* the considerations above had been taken into account. Of course, no one in their right mind would require that the gene should, moreover, programme its body to be capable of calculating degrees of relatedness before deciding how to distribute its largesse and make various sacrifices. That would be as absurd as saying that the genes that enable human beings to speak a language governed by a complicated system of grammatical rules should first equip them with an explicit theoretical knowledge of those rules – in which case even Chomsky's competence in English would be a mystery!

A simple recipe for kin-directed altruism would be Haldane's. A rare gene that disposed one to go to the rescue of two brothers at less than total risk to oneself should have a selective edge, for their combined descendants would have (among them) as many copies of that gene as would one's own. In general, if X is related to Y to degree R, we can calculate the circumstances under which X would behave altruistically towards Y.

Suppose X has the choice between helping Y or not so doing. Suppose P is the probability of total loss of individual fitness *for Y* if help is *not* forthcoming; and that q is the probability of such loss *for X* is help *is* extended. Then X 'should' help Y just in case p times R exceeds q ($q < pR$). In this very simple case we are, of course, also assuming that all that matters is whether to act or not to act, and that *if* one acts the outcome will be wholly successful – so that, in the case of my brother's drowning, if I dive in to save him, I shall most certainly succeed in doing so, but with risk p to myself. In general the factors of risk to both parties are of course somewhat more complicated, but the general idea remains the same.

We have spoken here of a rare allele 'for' kin-directed altruism. This is not, however, to suggest that kin-directed altruism is a monogenic trait. Of course it is not; but the value of any polygenic trait can be strongly influenced by single mutations in its basis. Thus the rare allele 'for' kin-directed altruism would have to be understood as functioning *with* the other (perhaps not at all rare) genes in the rest of the basis to make bodies that ensure the rare allele's future replication – and so the argument above would go through just as before. It would likewise go through even if two, three, four or more rare alleles were needed within a polygenic system to turn the bearer into a kin-directed altruist. Each of those rare alleles would still, as a 'best bet' to ensure future replication, make its contribution in such a way as to dispose the bearer to kin-directed altruism. All that would happen in this last case, however, is that the rare alleles would be re-assembled in bodies related by descent rather less frequently than before. But so long as there is genetic variability for kin directed altruism, as would be the case even here, it will respond to selection, weak though it may correspondingly be.

Thus once a gene or genes for kin-directed altruism arose, they would spread through the population at the expense of other alleles at the loci concerned. In due course the whole population would be one of kin-directed altruists, with the once rare allele(s) responsible for this trait no longer being rare, but rather predominating. And once in the driving seat, they are highly unlikely to budge, for reasons which will emerge below. In brief, kin directed altruism is an evolutionarily stable strategy; and it is to the explanation of this notion that we now turn.

2.2 Evolutionarily Stable Strategy

This notion was introduced by Maynard Smith and Price (1973). The best way to explain it is by a simple example. Suppose a group of organisms in times of food shortage forgoes reproduction to prevent overconsumption. This group has a better chance of survival than any other whose individuals do not so restrain themselves for the sake of their group. If, however, a mutation arises that produces an egoist, the picture is likely to change dramatically. The egoist will consume food at his fellows' expense and thereby – all other things being equal – be fitter. His descendants will contain the egoist's gene and will in turn outbreed the altruists. Thus the gene for egoism will spread through the population at the expense of competing alleles. (Of course, the group may as a result die out, but whether it does or not is irrelevant for our purposes.) We can safely assume a genetic basis for the earlier 'strategy' of altruistic restraint. But our reasoning shows that it is not stable in the evolutionary sense. As Dawkins (1976, p. 78) trenchantly puts it, an evolutionarily stable strategy 'is stable, not because it is particularly good for the individuals participating in it, but simply because it is immune to treachery from within'. In general a behavioural strategy is evolutionarily stable for a population if, should most members follow it, there is no alternative or mutant strategy that would make its followers fitter.

Now should a whole population become one of kin-directed altruists, that behaviour would be evolutionarily stable. It is quite clear that kin-directed altruism as we have described it is immune to the threat of new genes inclining their bearers either to total or to indiscriminate altruism. Kin-directed altruists would receive a net benefit from indiscriminate altruists, for they will always be taking from the latter without always giving; and total egoists would just not be doing the best they could to ensure the spread of egoist genes. For, to do that the gene for egoism would have to be a gene for kin-directed altruism, as explained above!

If members of a species always lived in kin groups, and physical separation on average correlated with degree of relatedness, then an allele could be as successful as possible by simply programming its bearers to help those close by. Under the assumed conditions this spatially determined altruism would be tantamount to kin-directed altruism, and only a cunning experiment could distinguish the two. The altruism would be extended without any need to recognise one's kin, consciously or subconsciously; all that would have to be taken into account is physical distance separating parties.

But in general, of course, although members of a brood or litter will spend some of their lives in close proximity, they will move around encountering both kin and non-kin. It is therefore interesting to enquire whether animals might not be able to discriminate their kin by sight, touch, sound, taste or smell in order to be altruistic with due discrimination. And indeed it has been found that frogs show some ability to distinguish close relatives from other strain members even though they had been isolated from each other before hatching (Blaustein and O'Hara, 1981).

We have explained above the basic ideas of kin selection and evolutionarily stable strategy. We were careful to define exactly what is meant by degree of relatedness and to explain why limited polymorphism is unimportant. We have also countered the misguided criticism of an allegedly naive 'one gene – one trait' correlation. Other misunderstandings of kin selection have been ably disposed of by Dawkins. There are, of course, still many practical difficulties in the way of measuring R between members of a wild population with an unknown degree of inbreeding and uncertainties of parenthood (Michod and Hamilton 1980).

Before turning to specific applications of sociobiological theory, a few general points about evolutionary explanation of the origin and function of different patterns of behaviour are in order.

A predator is hungry. It sees some prey. It stalks it, overcomes it in a final dash and settles down to a meal. Some of this behaviour might be instinctual, some of it learned. A *proximate* causal analysis would tell all there is to know about the development of motor skills through infancy, the maturation of closed programmes and the incorporation of past experience into open programmes. It would speak of the animal's internal motivational state at the time, characterising it, say, as one of hunger or of low blood sugar levels, depending on the level of analysis chosen. It would complete the explanation of this particular historical episode by mentioning all the relevant environmental stimuli that finally triggered the predator's actions – the shape, colour and size of the prey, its odour, etc.

This, however, is not the whole story of how the predator chose that prey and acted in that way. An *ultimate* causal analysis would specify further the respects in which the behaviour displayed and presumably inherited from forebears had been adaptive in the environments in which they found themselves. Suppose the present predation takes place in a completely new environment (as far as that species is concerned). Suppose the prey animal is a very uncharacteristic choice. Then the behaviour sequence that the predator executes may be quite 'out of character' but still highly successful. In such a case it will be much more difficult to bring the ultimate causal analysis into contact with the proximate one. When the behaviour to be explained, however, is characteristic of the species and takes place in an environment similar to that of its forebears, the ultimate causal analysis perfectly complements the proximate one. The ultimate causal analysis will explain why genes for that behaviour were there; and the proximate analysis will explain how, given that the genes were there, the behaviour arose.

Psychology and sociology provide proximate explanations of human behaviour; and ethology to a large extent gives proximate explanations of animal behaviour. The aim of sociobiology is to complement these with ultimate explanations. Compare Ghiselin (1974, p. 253): 'It is not necessarily wrong, but it may be uninformative, to say that an animal does something "for pleasure" (or in what passes for metaphysically untainted terms, "because it is rewarding"). We want to know why pleasure accompanies the act.' Ghiselin is here pointing out the inadequacy of hedonistic psychology's proximate explanations of behaviour. He makes the same point later in the same book (pp. 218-9). Darwin (1874, p. 135) was quick to point out that moral philosophers had been wont to explain social behaviour in terms of psychological impulses – such as the desire for pleasure. He maintained that all animals act more or less impulsively. Sentiments might come into play, but their role would be that of a proximate mechanism, having a more fundamental significance in protecting survival and reproduction. It might be that sentiments are a necessary condition for the existence of a society but, for not mentioning selective advantage, the explanation was not sufficient. For this reason, psychology could take but a limited role in the scientific explanation of social phenomena. This point has been most inadequately grasped even down to our own time; indeed, the oversight may be regarded as the basic reason why a truly evolutionary psychology has yet to be founded.

Thus in the human case in particular it is important to know the extent to which environmental influences at present shaping ontogeny are like those that shaped it in the past. There are two main reasons for thinking that the similarity is strong enough to bring sociobiology's ultimate causal explanations of human behaviour into illuminating contact with proximate explanations. The first is that the most profound shapers of behavioural traits are those of family life in infancy and early childhood. These shapers have themselves not been wholly disfigured by past and even recent cultural history. The second is that genetic adaptation, as claimed above, comes quickly in the wake of cultural change. It thereby helps close the gap between supposedly hunter-gatherer genes and space age environments. Understandably, social scientists have taken a protectionist stance. Thus Kunkel (1977, cited in Losco 1981):

> *That* one learns, and *how*, are matters of biology and vary little among populations. But *what* one learns, and *when*, being matters of culture and history, are the proper domains of sociology. The explanation and prediction of any specific behaviour, then, cannot be done without a sociological analysis.

But, of course, sociobiology will not render proximate explanations superfluous. It does, however, provoke new questions about social behaviour (in other species as well as our own) which otherwise might never have been asked or answered (Barlow, 1980, cited in Losco, 1981).

2.3 Applications of Sociobiology

In his book *Animal Dispersion in Relation to Social Behaviour*, Wynne-Edwards (1962) maintained that a large part of social behaviour was a result of group selection. In particular this would hold for any altruistic constraint on individual breeding in the interest of group survival. But as we described in chapter I, group selectionist explanations cannot be seen as alternatives to allele selection, only as alternatives to individual selection. Both groups and individuals are specific environments – though environments that alleles helped to shape in a more significant way than any physical environment – in which alleles compete with other alleles on the same locus for maximum representation in the following generation.

We must emphasise that kin selection is only a synonym of individual selection. The term stems from the rather belated recognition that alleles identical by descent to one's own are not only to be found in one's offspring but also in one's kin. Kin groups pose no special problems for alleles adapted to an existence in individuals. They can still go for all out reproductive success – even if for the individuals concerned this might occasionally mean that they have no own offspring. In non-kin groups selection may favour alleles (not individuals as in the case of kin selection) which do not maximise their own representation in the group they find themselves in. This can, of course, happen only because their representation in the whole species, or any other wider collections of groups, is thereby increased.

In this chapter we intend to survey several areas of application in which sociobiology has achieved a measure of success, albeit controversial. We shall start with an analysis of aggression and submission, showing how the notion of an evolutionarily stable strategy can be fruitfully applied. We cannot here undertake a full review of all the evidence for and against the methods and theories of sociobiology in each of its diverse areas of application. For this the reader is referred to several detailed studies (Chagnon and Irons, 1979; Ruse, 1979a; Markl, 1980; Alexander and Tinkle, 1981). Our intention is only to impart their flavour and situate the subject in order to get to grips with the question of how biology connects with moral philosophy.

Aggression

Consider a population consisting entirely of individuals who, upon meeting in competitive circumstances, fight until one dies no matter what the risks of injury to the victor. Let us call them hawks. By contrast a dove is one who flees all such encounters and never initiates them. In a population of hawks a dove will enjoy an advantage if the chances of escape are usually good. While the hawks maim and kill one another, the dove can get on with his feeding and breeding. The proportion of doves thus grows. (Both here and throughout we are assuming the behavioural traits in question to be heritable – an assumption shown in the first chapter to be fully justified.) But when the proportion of doves exceeds a certain threshold, the hawks obtain a new advantage: they are fighting less frequently, because there are fewer hawks around, so they can monopolise resources with less risk of injury or death. The evolutionary pendulum swings back in their favour.

The elegant game theoretical analysis of Maynard Smith and his co-workers (1973; 1976) shows how a stable equilibrium will be reached at a certain ratio of hawks to doves. The precise numerical value of this ratio depends on the evolutionary costs and benefits of injury, time-wasting and so on. Their quantification obviously faces considerable methodological difficulties. What the game theoretic approach sets out to show, however, is the qualitative character of an evolutionarily stable strategy; and this can be done without exact values of the various coefficients. The stable ratio of hawks to doves will probably be approached by damped oscillation. It need not be one of population numbers of consistent hawks and consistent doves. Stability can be obtained also by the appropriate admixture, in each member, of hawkishness and dovishness. One can also consider other strategies with such suggestive names as bully, retaliator, and prober-retaliator, and determine the stable ratio of followers, or admixture of inclinations in each member in the population.

Frequently in evolutionary thinking there are two aspects to the explanation of a phenomenon. The first is the genetic stability or entrenchment of a trait. The second is the trait's evolutionary origin. The latter is always the less tractable. Maynard Smith's game theoretic methods help us to understand the former.

Even with morphological features it can be difficult to reconstruct the exact sequence of steps, and their genetic origins, whereby the species acquired the feature. This is so even though the paleontological record provides evidence of morphological evolution. Unfortunately in the case of social behaviour there is little or no paleontological record. This does not prevent us, however, from recognising some theories of behaviour as much more plausible than others, when judging them with an eye to the ultimate causal analysis that they provide. Let us compare, for example, Lorenz's proximate theory of aggression with the ultimate one of Maynard Smith.

Lorenz's major thesis (1963) was that there is an aggressive drive in each individual which builds up steadily and must eventually find expression. If no opportunity presents itself the 'action specific energy' will be released upon little or no conventional stimulus to aggression. Experiments on fish in Lorenz's own laboratory have in one case confirmed his 'drive discharge' model (Rasa, 1971). Withdrawing aggressive stimuli through isolation led to an increase of aggressive behaviour on the part of males when they were reunited with conspecifics.

The males even attacked females. But similar experiments by Heiligenberg and Kramer (1972) on a different species of fish failed to reproduce these results. See Franck (1976) for a good discussion of this controversial field.

Now, for a sociobiologist the dispositions postulated by the drive discharge model are *a priori* unlikely to have evolved, because they could so easily become maladaptive. An animal running the risk of injury to itself through gratuitous acts of aggression only to release pent-up aggressive 'energy' would be a highly imperfect survival machine. Of course, what we may be observing in the aggression of individuals returning from isolation is a pathological behavioural product of unnatural environmental circumstances. But then the special interest of Lorenz's account attaches only to such pathologies; and he still owes us an account of the aggressive behaviour an animal is disposed to display in the normal conditions that obtain during its phylogeny. This is exactly what sociobiologists are trying to do.

It is worth noting that ethology was conceived as the Darwinian study of animal behaviour, in complete analogy with traditional comparative morphology or anatomy. The latter studies homologous traits and reconstructs the lines of evolutionary descent in search of a deeper understanding of the structure and functions of various organs. Lorenzian ethology, however, was based on an unsophisticated Darwinism that lacked any knowledge of population genetics.

Wilson (1978) in *On Human Nature* cites a survey by the anthropologist Sipes (1973). This revealed that in ten notably warlike human societies and ten pacific societies, the former display a generalised heightening of all forms of culturally expressed aggression. This is apparently at odds with what one might expect if the drive-discharge model were correct; a conclusion that is valid, however, only if we assume that the warlike peoples did not produce significantly more action specific energy. This is a general methodological shortcoming in common with all such conclusions based on demographic data: that a perhaps unwarranted assumption of biological identity of the different geographical populations is being made. That is to say, it is being assumed that the different populations have identical means of aggressiveness in identical environments. If this is not so, then Lorenz might, of course, still be right despite Sipes's observations. As we have noted in chapter I, many human sub-groups have been reproductively isolated under different conditions for long enough to exhibit genetic diversity as a result of different selective pressures and different histories of mutation.

Evidence can be brought to bear, however, on Lorenz's thesis in a more decisive way from psychological experiments. People who see films with aggressive content afterwards incline markedly to aggressive acts and statements of aggressive intent when presented, for example, with hypothetical choices of punishments for various misdoings by others. It appears, however, that if one is already in an aggressive mood then seeing such films, or being able to administer 'punishments' can reduce at least the physiological symptoms of this mood (see discussion in Eibl-Eibesfeldt, 1967). The implication thus seems to be that one needs specific stimuli both to get into an aggressive frame of mind and to act aggressively once in it. There does not appear to be any aggressive 'drive' that simply builds up in the absence of such stimuli. The multifactorial nature of aggression and the empirical inadequacy of the drive-discharge model is established in the excellent discussion given by Hinde (1974) in *Biological Bases of Human Social Behaviour*, written before the advent of sociobiology. As just mentioned, the evolution of a drive-discharge mechanism is unlikely. Secondly, the evolutionarily stable strategy approach allows an account of aggressive behaviour following various other strategies: appeasement, submission, retaliation, etc. In marshalling all the data on human aggression, sociobiologists would see their task as that of explaining the evolution of a mechanism whereby in different environments people would learn different kinds of aggressive behaviour keyed to stimuli characteristic of those environments. It seems obvious that diverse forms of inter-group aggression, from football games through lynching parties to nuclear warfare are not in any way genetically pre-programmed but are culturally acquired. The basic tendency towards such behaviour is, however, a product of evolution and must be studied as such. Here we might ask whether the emerging data on organised warfare and cannibalism among chimpanzees (Goodall, 1979) could be interpreted as showing that evolutionary considerations are relevant to the study of human warfare.

The function of diverse forms of aggression must be elucidated in terms of evolutionary advantage to the genes of aggressive individuals. In this connection, we mention in passing Wilson's classification (1975) of aggression into eight broad types: territorial, dominance, sexual, parental disciplinary, weaning, moralistic, predatory and anti-predatory. The evolutionary function of some of these is obvious to a non-biologist. Others, particularly the moralistic type, are much more intricate. One thing is clear: aggression is a highly adaptive trait, and it might very well be ill-advised to seek its elimination in a socially engineered utopia; even if that did not amount to the wrong of trying to achieve the impossible. Many forms of human aggression are inextricably linked with group identity and group cohesion. These in turn determine a degree of assortative mating important for maintaining that genetic variability which represents mankind's potential to meet the evolutionary challenges of the future. We are not saying that the evolutionary function of aggression was to maintain variability of the population, but rather that genetic variability could be regarded as a beneficial side effect of the evolution of aggression. Reformism or utopianism should not take biology lightly. There are some channels into which the tide of human history cannot be swept; and there are evolutionary cul de sacs we must navigate past.

Mother's Brother

For a long time anthropologists had noticed in the case of several cultures how important a mother's brothers were in the rearing of her children. Frequently these men are in positions of greater trust and responsibility than the biological fathers. Anthropology provided no ultimate causal explanation for this arrangement, concerned as it was more with observation and classification. Sociobiologists noted that in just these societies knowledge of paternity was significantly less reliable than in others. It is easy to see that if the probability of the husband's being the father is below a certain value then the mother's brother is on average more closely related to the child than the husband. Assume there is not too much inbreeding. If the probability that any child's biological father is its marital father is less than 1/4, then a mother's brother is on average more closely related to the child.

Depending on the level of promiscuity, the degree of relatedness between uncle and child lies between 1/8 and 1/4, since the uncle and the mother have the same mother. It follows from considerations of inclusive fitness that mother's brothers should take on the parental role.

All that this sociobiological explanation accounts for is the genetic stability of the promiscuity-cum-uncle guardianship arrangement. We cannot simply assume that the guardianship pattern evolved against an already established backdrop of paternal promiscuity or vice versa. There are obvious necessary conditions for the origin of both these features of social behaviour. It must have been possible for both the genetic tendency to promiscuity and the genetic tendency to care for sister's offspring to increase by virtue of the relative unimportance, within a group life, of the more intensive care that a father is likely to give to his own children rather than to his sister's.

Discriminating Reasons for Less Discriminating Altruism

Some allegedly recalcitrant anthropological data have been advanced by Sahlins (1976) in his thin pamphlet against sociobiology. He claims that Polynesian kinship systems contradict sociobiological kinship theory. Polynesians have extremely flexible adoption arrangements: 'Polynesia generally is famous for adoption practices that violate the moral logic of kin selection with regard to parental care, concern for one's own offspring as against those of genetic competitors etc.' (p. 48). As Freedman (1980) has pointed out, however, inbreeding on these islands is very high and any two people are related as closely as are siblings in an outbreeding population.

The Polynesian adoption system can thus be explained by sociobiological considerations provided we add to Freedman's observations the following qualification. The most successful kind of rare allele is that which enables its bearers both to recognise its presence in other bearers, and then to act duly altruistically towards them (Dawkins's Green Beard Effect (1976)). If such an allele spreads to fixity in any population – as it is highly likely to do – then the altruism that results is tantamount to universal altruism. In a population of green beard altruists, altruism towards green bearded members may not be the same in principle as universal altruism, but happens to amount to the same. Now, once such background sociality has been established, the spread of alleles for effects similar to the green beard effect is radically changed. Consider an allele producing, say, blue ears, and the tendency also to help blue eared fellows rather than ones without blue ears. Blue ears will benefit from the altruism of green beards, but not vice versa. Thus we would expect the allele for blue ears to spread to fixity. Capitalising thus on green beard magnanimity, however, is not the exclusive prerogative of blue ears. Mutant alleles for any number of 'altruism based on recognition'-effects would get in on the act. The result would be a population of altruists whose altruism is like the vector sum of manifold altruistic inclinations along various phenotypic co-ordinates. An arbitrary member of the population may thus be expected to distribute his altruism preferentially by degree of phenotypic resemblance. And now in a highly inbreeding population, this would be tantamount to universal altruism.

We must emphasise that this is the evolutionary picture of the genetically based altruism among Polynesians that must hold good if sociobiology is to accommodate the observed distribution of altruism without regard for *genealogical* kin. The only other possibility is the one that Freedman must have had in mind as the only one. This is, that altruism began with an initially rare allele that programmed its bearers with an indirect rule for directing their benevolence: such as to help those in the same hut or village. As matters turn out, the recipients would tend to be genealogical kin. By kin selection the allele for the indirect method could spread to fixity. Subsequently, dislocation of social patterns – such as geographical restructuring of characteristic dwelling patterns – could bring it about that the indirect rule 'misfired' in the sense that recipients were by and large not (close) genealogical kin. The indirect behavioural rule, however, embellished by social custom, would nevertheless prosper, and even hold its own against any more discriminating indirect rule precisely because the population was so inbred. What Freedman crucially overlooks is that any indirect rule is vulnerable to invasion by a rule based on recognition (green beard) effects. Thus it is essential to provide the account of the previous paragraph to show how to take care of this effect while nevertheless saving Polynesian appearances in a highly inbred population.

In general, as Cattell (1972) has pointed out, anthropologists have been far too impressed by superficial dissimilarities among societies they study. What is important is not why this society has such-and-such a kinship system while that society has such-and-such another, but why they have kinship systems at all. This, of course, is the main thrust of sociobiology, and it need not be considered as conflicting with those social sciences mainly concerned with the delineation of points of difference. This interest in surface manifestations, however, should not be allowed to issue in rash quantifications of the degree of cultural variability across the globe. It is difficult to specify even the dimensions along which this variability is supposed to range, let alone mark off units on the axis concerned. Moreover variabilities in such eminently quantifiable features as mass and length are strictly speaking incommensurable. While natural scientists may from time to time massage their data, nevertheless social scientists appear to be unique in having little in the way of data to massage, but a great inclination to indulge in the afterglow. A good example of this kind of softheadedness is Ruyle's claim (1977) that 'The gulf between humanity and other creatures . . . is as great as the gulf between life and nonlife, and any theory of cultural evolution which fails to grasp this is doomed to failure.'

Parent-offspring Conflict

Trivers (1974) has used the notion of inclusive fitness to shed new light on the interactions between parents and their offspring.

Take any species in which females can produce more than one offspring. Suppose the offspring depend on their mothers for at least some period of their lives. Then on purely theoretical grounds one can expect some degree of parent-offspring conflict.

In a typically mammalian species the mother provides food and protection for her young; both she and the offspring benefit genetically thereby. There comes a time, however, when the mother would be diminishing her inclusive fitness were she to maintain this arrangement. The growing offspring is now better able to fend for itself. The mother should, in her own genetic interest, stop investing in this child and start devoting her resources to potential future offspring. Depending on the degree of relatedness between siblings, there comes a point where mother and offspring 'agree' to end the dependency. But the agreement usually emerges only after a period of conflict. Conflicts over the onset and rate of weaning can be observed in practically all mammalian species. Why is this so? Let us assume a faithfully monogamous species. Call the offspring Ego. Ego is related to his mother by 1/2 and to his own children (assuming outbreeding) by 1/2. He is likewise related to his siblings by 1/4 *but* – and here is the rub – to his nieces and nephews by 1/4. Thus from the point of view of Ego's inclusive fitness a unit of parental investment in him is worth twice as much as a unit invested in a sibling. From the point of view of Ego's mother's inclusive fitness, however, these units are of equal value. It follows that there will be a period of conflict between Ego and his mother over the allocation of her investment. When he is so self-sufficient that a unit denied him will detract from his own personal fitness less than half the extent to which that same unit, invested in a needier sibling would increase that of the latter, then considerations of inclusive fitness would enjoin him to acquiesce in maternal reallocation of care.

But before that time Ego can be expected to act selfishly in competing for the available benefits. Thus not only does genetic geometry locate conflict on the vertical axis in a way little appreciated before the advent of sociobiology; but it also explains the well-known conflicts on the horizontal axis that have earned the general label of sibling rivalry.

Theory predicts that the mother wishes her children to co-operate with one another more than they themselves do. Thus Trivers:

> If this argument is valid then it was clearly a mistake to view socialisation in humans (or in any sexually reproducing species) as only or even primarily a process of 'enculturation', a process by which parents teach offspring their culture . . . According to the theory presented here socialisation is a process by which parents attempt to mould each offspring in order to increase their own inclusive fitness, while each offspring is selected to resist some of the moulding and to attempt to mould the behaviour of its parents (and siblings) in order to increase its inclusive fitness. Conflicts during socialisation need not be viewed solely as conflict between the culture of the parent and the biology of the child; it can also be viewed as conflict between the biology of the parent and the biology of the child. (p. 260)

We might add a note on a special case of conflict: that of parental manipulation of progeny. In general, the mother has an equal interest in every one of her offspring. (But large fitness differences between siblings could lead the mother to treat her offspring differently, distributing her investment so as to maximise her own fitness – indeed, even increasing those previous fitness differences among her offspring.) In certain cases it may pay the mother, in terms of her inclusive fitness, to suppress the reproduction of one of her brood in order to enlist the home help of the offspring concerned. This hypothesis has been proposed to help explain human male homosexuality. For it is not only possible that genes for homosexuality prevail because of superior heterozygote fitness. It is also possible that parents can gain by 'manipulating' sons with that genetic tendency so as to make them non-reproductive helpers – as they are known in several other species – in raising their siblings.

Another interesting possibility is that parents may not only manipulate the behaviour of their offspring, but even 'manipulate' the sex ratio in their own favour. The two sexes have different variations in reproductive success. Women who bear children at all bear a number much closer to the average for their sex than is the case with males. Thus it is in the interests of a fit female mating with a fit male, to 'prefer' the likely fit outcome of that union to be male – a son who, because of his greater reproductive success, will increase the mother's inclusive fitness. There is evidence from primitive societies that in the higher ranks female infanticide is more frequent (Dickemann, 1979) and that high-ranking females have more male than female offspring.

This parental manipulation of the sex ratio predicted by Trivers and Willard (1973) can be observed in many animal species, and has been borne out by studies on South American indians, and stratified societies in China, India and Europe. But this is not all. Much more far-reaching conclusions from this single consideration have been advanced in a paper by Alexander *et al.* (1979). Phenomena to be explained in this way include various sexual dimorphisms, longer gestation periods for boys, higher mortality rates for males, longer suckling of males, more widespread female biased infanticide, more rapid senescence of males and several others besides. Freedman (1980, p. 15) points out that 75 per cent of the world's culture areas are polygynous, a fact which 'seems to point to this system as being most in tune with the species' natural requirements (fulfilling both the male and the female strategies)'. (p. 15)

Human Sexuality

Another more exotic outcome in the application of sociobiological theory to human sexuality is the following suggested explanation of why the female human being is unique in concealing ovulation (Alexander and Noonan, 1979). It is part of her reproductive strategy of keeping her mate near to her and her children. For, if ovulation was easily detectable, the male would only need to protect her from other males during that period. This way he would ensure that her offspring would be his as well; and for the rest of the time he could be trying to make merry with other females. Moreover, if the female gets a male who is highly fit but a cad as a father, she will not be too well off in bringing up her offspring. A compromise reproductive strategy would therefore be to mate with and keep a relatively less fit but more devoted father; and such males have a chance only if there is not a constant battle for oestrous females going on, with the fitter males constantly winning.

These and many other intriguing or outrageous explanations of quirks and regularities of human sexuality can be found in several recent publications in the burgeoning paradigm. For example, Symons (1979) has pressed the hypothesis that the sperm's being tinier than the egg would lead to radically different reproductive strategies for the two sexes. He uses it to explain the following seven contentions about human beings:

1 Intrasexual competition is more intense among males.
2 Men incline more to polygamy than women do.
3 Men are more sexually jealous than women.
4 Men are more easily visually aroused.
5 For women, being physically attractive is more important; for men, prowess.
6 Men more than women like sexual variety for its own sake.
7 Copulation is considered a service or favour that women render to men.

Symons also sees the female orgasm as a side effect of biological evolution, rather than as serving an adaptive function. And in taking issue with the radical feminist Sherfey (1972), he claims that 'The sexually insatiable woman is to be found primarily, if not exclusively, in the ideology of feminism, the hopes of boys, and the fears of men' (p. 92).

Ghiselin (1974) had a few years earlier made so bold as to extrapolate from the reproductive strategies he had laboriously and painstakingly exposed in 'nether' species. In a book admittedly devoted more to the sexual proclivities of bladderworts and crabs, and containing only the odd reference to our own species, he concluded:

> Man's brain, like the rest of him, may be looked upon as a bundle of adaptations. But what it is adapted to has never been self-evident. We are anything but a mechanism set up to perceive the truth for its own sake. Rather, we have evolved a nervous system that acts in the

interests of our gonads, and one attuned to the demands of reproductive competition. If fools are more prolific than wise men, then to that degree folly will be favoured by natural selection. And if ignorance aids in obtaining a mate, then men and women will tend to be ignorant.

Altruism

Human sociobiology is a belated embellishment of T. H. Huxley's remark (1947): 'I see no reason to doubt that, at its origin, human society was as much a product of organic necessity as that of the bees.' Wilson's own earlier work was indeed on insect societies (1971). Just as a biologist will ask after the survival value of pronounced morphological features of an organism, so the sociobiologist asks after the evolutionary causes of man's most important dispositions to social behaviour. If the theory of natural selection can account for these then we have the beginnings of a possible ontological reduction in the social sciences. Any theory with anything to say about the evolution of social behaviour in human beings must address itself to the question of altruism, which is the bedrock of social behaviour.

The forms of biological altruism defined by sociobiologists bear at best a troubled relationship to the altruism of everyday moral parlance. The latter importantly involves consideration of the agent's sincere and conscious intention, his lack of personal self interest (a very different matter from his biological inclusive fitness) and, also, the non-violation of rights of other persons. Sociobiology has co-opted the term 'altruism' from everyday language and pressed it into newly defined technical service for evolutionary theorising. Obviously its usefulness in this new role depends importantly on the extent to which biological explanations using it can be brought to bear on a significant portion of human social behaviour. In doing so, we are only following the inevitable course of scientific concept formation and theorising. For who knows, the new term may come to displace the old even in everyday language as the new science seeps into popular consciousness. The everyday notion could come to be seen as a metaphorical extension of the scientific one, in much the same way that psychic energy in a final analysis might have to be grounded in the physicist's notion of energy, even though the word 'energy' might have come to be used first and foremost in connection with people's drives and capacity for work. As Lorenz (1973) has noted also, the word 'etiology' now used exclusively for the causal pedigrees revealed by science has its etymological origin in a Greek root denoting human intention.

Biologists today should define altruism as follows: an individual is said to be acting altruistically towards others if in so doing he appears at the time of the act to be decreasing his own chances of survival and reproduction but increasing the others'. By so defining their notion of altruism – a notion already not necessarily in exact agreement with the everyday one – they can identify and correctly describe certain kinds of behaviour as examples of *kin-directed altruism* or *reciprocal altruism* (to be defined below). They can do so even though the ultimate evolutionary explanations given for the origin and stability of such behaviour shows that it is behaviour promoting the interests of the agent's genes. To be sure, much of what passes in ordinary language as instances of altruism will also fall under the heading of kin-directed altruism or reciprocal altruism. But it will turn out to be altruism at best from the standpoint of individuals, rather than from that of their genes.

But what about saints and heroes? Is there any residue of characteristically human behaviour that can be called genuinely disinterested altruism in the everyday sense; that cannot be explained by appeal to any genetic mechanism for its perpetuation; and yet – most important of all – is stable and prevalent enough in our species to call for special recognition, confirmation and explanation from the biological and human sciences? For the sociobiologist fullblown altruism of this sort is pure existential dogma. It just cannot occur often enough to be a species characteristic calling for ultimate causal explanation. To be sure, some acts of reciprocal altruism might on occasion save genetically unrelated individuals. They might occur at great risk to the rescuer. It might be possible that no possible benefit accrues to him, either as the result of subsequent fame and social reward or as the result of the beneficiaries' gratitude and inclination to reciprocate in the future. But then these occasional acts can be explained as 'misfirings' of the disposition to behaviour of the broadly 'reciprocal altruist' kind, behaviour which in the evolutionary past of the species did not have the rare and laudable qualities of the 'misfiring' just described.

We have already discussed kin directed altruism in the second section of this chapter, explaining how the mechanism of kin selection can favour heroic inclinations when the beneficiaries are closely related to the hero. Less dramatic examples than that of leaping into a torrent to save two brothers are all around us, especially in parental and grandparental care. It is hardly necessary to adduce psychological or sociological studies to attest to the extent of preferential treatment of one's kin. More interesting light is shed by studies that reveal the finer discriminations in line with kin selection theory. For example, Sebastian (cf. Freedman, 1980) questioned subjects about their readiness to steal an otherwise unavailable medicine for someone suffering from an illness. The more closely the protagonists were related, the more readily the subjects said they would condone this stealing. Ginsburg (cf. Freedman, 1980) has discovered that the kin bias of altruistic intent develops through childhood in a Piagetian fashion. Grandparents remain faithful to the dictates of kinship. Only in the case where their own children were beyond reproductive age would they prefer to help their grandchildren. Perhaps the most extreme example that may fall within the scope of kin-directed altruism is the self-sacrificing behaviour of kamikaze pilots. The modern Japanese population is descended from a few thousand Korean invaders; being therefore more closely related genetically than Europeans and Americans, say, their martial excesses are perhaps more readily understandable. (See other examples of Japanese social cohesion (particularly adoption arrangements, like those of the Polynesians) in Freedman's *Human Sociobiology* (1980).)

Reciprocal altruism is a more subtle matter genetically, even though a vast amount of everyday give and take may be due to it. It is the type of altruism where the altruist can reasonably expect reciprocation from those he helps. Trivers (1971) was the first to explain how it could evolve through gene selection alone, in a population whose members can recognise and remember each other, can bear grudges and remember debts. A recent paper by Axelrod and Hamilton (1981) has provided a sound mathematical basis for reciprocation theory. All one need assume is a certain probability that interacting individuals will interact again in the future. Each interaction can be understood, in an idealised fashion, as a round of the game called Prisoner's Dilemma. There are two moves available for each player – co-operation and defection. If I co-operate but my opponent defects I get the sucker's payoff S and he gets the devil's payoff D. If we both defect we each get punishment P; if we both co-operate, we each get reward R. The catch of this particular game is the relative magnitude of these payoffs: $D > R > P > S$, with $R > (S + D)/2$. If the game is to be played only once, the strategy of defection can be shown to be evolutionarily stable. But if members of the population are to meet again, and can 'remember' past rounds with their opponents, then more interesting strategic possibilities arise. When this likelihood exceeds both (D–R)/(D–P) and (D–R)/(R–S), a strategy called tit-for-tat is evolutionarily stable. This strategy is very simple: you co-operate in the first round, and then in succeeding rounds do whatever the other player did in the preceding one. The main problem now is to explain how such a strategy could come to be applied by sufficiently many members of the population to guarantee its stability (for obviously if only a few members follow this strategy, it does not pay). Hamilton and Axelrod's line of reasoning is as follows: it pays closely related individuals to co-operate in prisoner's dilemma-like situations from the point of view of kin selection theory. Thus when a co-operative choice has been made the simple fact that it is reciprocated becomes a clue to relatedness. Thus 'co-operation' can spread into circumstances of less and less relatedness. Finally when the probability of two individuals' meeting each other again is sufficiently high, co-operation based on reciprocity can thrive and be evolutionarily stable, in a population of no relatedness at all.

Let us suppose now that social intercourse is tit-for-tatish. One would expect that ways would then evolve of ensuring that cheats are identified. It is extremely useful to one, when first meeting a new player, to know from others whether he usually takes and seldom gives; for then one can avoid being made a sucker. Our whole machinery of contract law, and the powerful social forces of good name and gossip, can be seen as cultural enablers of reciprocal altruism. Indeed some legal systems have sanctions maintaining reciprocal altruism. In West Germany one is under a legal obligation to go to the aid of a drowning person.

It should be added that reciprocal altruism does not rely exclusively on individuals' being able, consciously or unconsciously, to reidentify one another. It is also possible for it to give rise to what is known as inter-species mutualism. Obviously an individual can help or be helped by several others simultaneously, with their inclusive fitnesses being decreased. This, however, can only be the case if they are not members of the ingroup of the altruist and his beneficiaries. Who is a member of the ingroup depends on the selective mechanism by which this behaviour has evolved. Both kin-directed and reciprocal altruism evolve through gene selection. We wish now to consider a third possibility; could some form of in-group altruism have resulted from group selection? In-group altruism involves an average increase in the fitnesses of other members, with their genetic relatedness to the altruist playing no role, and with the altruist himself not expecting reciprocation. It is most frequently invoked whenever it is observed that animals seem not to realise their full reproductive potential.

Wade's experiment (1977) on Tribolium, the flour beetle, showed that what he called group selection can decrease reproductive activity. But whether this decrease is due to altruistic 'restraint' or is simply a matter of the beetles' being reproductively stunted, is unclear. If the former, then the altruism of those beetles cannibalising their young in order to keep population numbers down is a far cry indeed from the everyday notion.

As we remarked above, altruism in the biological sense can cover perhaps all prevalent cases of altruism in the everyday sense. Now we have just seen that it can cover much more besides. But in those cases where the notions coincide – as in, say, saving someone from drowning or from fire – we can see that the biological notion, embedded in the evolutionary account, provides a redescription of the altruism in terms of genetic self interest. This may disturb moralists, and probably caused much of the public furore about sociobiology. We would argue, however, that recognising genetic causes of behaviour does as little to demote everyday altruism as an indispensable category of moral appraisal as it does to banish free will from the theory of human action. We have speculated about the technical term supplanting the everyday one. But we may nevertheless ask whether, were sociobiological theory to be widely accepted, it would change the moral labels that we attach to good deeds and bad. Perhaps the whole system of moral appraisal enmeshed with folk psychology is integral to the efficacy of the very behavioural dispositions that the sociobiologists themselves are concerned with. This system provides the scale according to which reciprocation, reward and restitution is regulated. All that may change, however, is the authority of 'revealed' values: or rather the authority of their revelation. The origin and continuing importance of these values will become objects of closer scientific investigation.

Is there such a thing as pure genuine wholly disinterested altruism? The following are at least necessary conditions for an act to count as such. They serve to exclude the possibility of any hidden genetic cause that would cast the act in the light of genetic self-interest.

1 The altruism must be extended to non-kin.
2 The altruist must not expect reciprocation or reward from anyone learning of his action.
3 The altruist must not expect to lose his own entitlement to altruism

from others, in case he withholds help.

4 The disposition to this type of behaviour could not have evolved through group selection.

Briefly (1) says – 'no kin selection'. (2) and (3) say that there should be no positive or negative inducement, hence, 'no gene selection for reciprocal altruism'. (3) says – 'no group selection'. Expectation is to be understood in (2) and (3), not in terms of the agent's beliefs, but rather in terms of the biologist's assessment of consequences for survival and reproduction on the part of the altruist. This is of crucial importance, given the possibility of self-deception and the possibility of unconscious motives.

Biological necessities frequently find cultural expression that obscures or conceals their biological origin. An obvious example would be prohibition of in-group murder and of incest, backed by divine sanction. We specify *in-group* murder because *out-group* murder is frequently committed in the name of the same deity. This was so not only in the time of the Crusades. It is evident from the recently published sermons of Christian ministers in Nazi Germany. No doubt there are other examples (Iran and Northern Ireland), and jehads yet to come.

Let us try now to add some more detail to our imagined case of pure disinterested altruism. The altruist must help only members of an out-group. He must not help, even indirectly, himself, his kin, or any members of his own in-group. He must perhaps travel far away to foreign lands to do this; but he may not, as is arguably the case with missionaries, be a subtle arm of his own group's imperialism. He must contribute so marginally to foreign individual or group fitness, and do this at such high cost to himself, that such behaviour will be ignored by all selective forces (including group selection) even if it becomes more prevalent. This last condition is important because it is probable that natural selection has endowed people with a certain amount of altruism towards the people around them, who were no doubt mainly kin. People were probably only rarely transferred into another unrelated group in our evolutionary past. We are therefore unlikely to follow genetic maxims in adjusting our behaviour to life in a foreign group. In other words, altruistic dispositions are probably coded genetically as '*de dicto*' rather than '*de re*'. So only if the kind of altruism we are trying to envisage could not have evolved in the in-group situation, can we count it as non-biological altruism in the out-group situation. Obviously low benefit to the recipient and high cost to the altruist would have prevented selection favouring the behaviour concerned.

We are now in a position to make our imaginary case concrete. Someone travels to a distant foreign land of plenty, burning all bridges behind him. He proceeds to give away every morsel of food he finds or receives, to the near negligible advantage of the recipients. He abstains from sex in order to distribute food more effectively. Finally, no conceivable advantage accrues to his home group as a result of his project.

The absurdity of these conditions, despite the existence of a few 'living saints', counts strongly in favour of the adequacy of the scope of biological altruism. We have been assuming that biologically senseless altruism could not have been selected for – not even by group selection. It may be, however, that those who maintain that 'pure disinterested' altruism exists would be content with claiming that it evolves by group selection rather than gene selection. In this way, it might be thought, genetic self-interest can be excluded, and the altruist, though a product of evolution, does not further his own interest – that is, increase his own inclusive fitness – by his deeds. But this line of thought is mistaken. For a group selected altruist *does* increase his fitness in comparison with *non-members* of his group. Only in comparison with fellow group members is he, genetically speaking, unselfish or non-egoistic. So even group-selected altruists are genetic egoists in the context of their species.

Mother Theresa, born in Yugoslavia of Albanian parents and working among the poor of Calcutta, might count as a pure disinterested altruist. But supposing that, by her interventions, some of those helped have been able to breed who would not otherwise have been able to, or that in any other way her help has been more than 'minimal' in the sense specified above, then her *de dicto* altruism may well be of a kind that could have resulted from both kin and group selection. This would especially be the case had she already been subject to the kind of parental manipulation discussed above.

The biologist's final resort in the face of heart-rendingly pure altruism is to count it as too rare to have evolved by selective force. Saintliness prevails because saints cannot be prevalent. Its occurrence then would be put down to rare genotypic constitution or misfiring in unusual environments. Even with a cultural Überbau not all individuals can be expected to have identical altruistic dispositions. We expect some distribution of such dispositions, as with any other continuously varying phenotypical traits. One extreme of this distribution will always seem 'unbiologically' altruistic. But such cases would not be prevalent enough to warrant more than a pathologist's interest. It may call for proximate explanation, but it would certainly not call for ultimate evolutionary explanation.

In Hardin's words (1977, p. 5):

> Does pure altruism exist? We grant that any behaviour, no matter how remarkable, must occur now and then. Mere existence is not the important question: what matters is persistence. Can altruistic behaviour persist and increase, perhaps ultimately becoming the norm to which egocentric behaviour is the exception? Such is the dream of compassionate people. Is it a realistic dream?

Singer (1981) has conjured up a situation that he believes would make the evolution of genuine altruism more likely. In the Prisoner's Dilemma the best strategy is clearly not to be selfish. So evolution might favour those individuals in such dilemmas who are 'genuinely altruistic instead of making reciprocal exchanges on the basis of calculated self-interest' (Singer, p. 47). But could situations schematically similar enough have arisen often enough in our evolutionary past? He gives the following example:

> Suppose two early human beings are attacked by a sabertooth tiger. If both flee, one will be picked off by the cat; if both stand their ground, there is a very good chance that they can fight the cat off; if one flees and the other stands and fights, the fugitive will escape and the fighter will be killed. Here the odds are sufficiently like those in the Prisoner's Dilemma to produce a similar result.

But such situations were probably too rare to have had any effect on our genetic predispositions to help or defect under such circumstances. Normally one of the two 'prisoners' would always be better off. This would be the case if the predator or enemy had only spotted his partner; or if he were much better able than his partner to flee; or were a much faster runner, better climber, etc. If it usually happened that the one better able to escape nevertheless came to the aid of his partner, only kin and group selection, or selection for reciprocity could have brought this about – not, as Singer believes, a specific adaptation to the odds of the Prisoner's Dilemma game. Singer correctly concedes that since we are able to communicate, possibilities for reciprocal altruism have increased enormously:

> Reciprocal altruism may be especially important within a group of beings who can reason and communicate as humans can, for then it can spread from a bilateral to a multilateral relationship. If I help you, but you do not help me, I can of course cease to help you in the future. If I can talk, however, I can do more. I can tell everybody else in the group what sort of person you are. They may then also be less likely to help in the future. Conversely, the fact that someone is a reliable reciprocator may also become generally known, and make others readier to help that person.

Elsewhere, Singer also shows a certain lack of biological insight that leads him to unwarranted conclusions. For example, he asserts that genuine altruism exists. His only evidence is that for thirty years more than a million Britons a year had donated blood to the National Blood Transfusion Service without being paid or otherwise rewarded for their donation. A sociobiologist would wish to observe these people after their donation to see whether they tell their friends and relations of their deed, thereby enhancing their reputations as altruists. This could be reward in itself. Others who learn of the donation might be inclined to reward the benefactor. Furthermore a biologist might maintain that thirty years is not a long time, and that such behaviour might be an atavistic form of kin-directed altruism that has not yet been successfully counter-selected. Certainly there are social pressures to make people donate. Singer's argument rests on the assumption that rationality can overcome emotions or the genes. One of his examples is the fact that people take the pill even though this is not in the interest of their genes. But people have not been taking the pill for very long. The living conditions in those countries where the pill is taken by a sizeable fraction of the female population favour k-selection; a few well-looked after children hold more promise than do large broods, who would be less well cared for.

Here sociobiologists prefer a hard-headed and perhaps cynical look at the available evidence. E. O. Wilson sees our behavioural repertoire as not very different from that of any other diploid social species. Man, like lions and chimpanzees, kills and eats conspecifics under certain conditions. For the most part, he conducts himself in accordance with a set of implicit values shaped by evolution for his survival. Evolutionary biology will make us reluctant to accept exaggerated claims for man's moral values in so far as these are to be submitted to science for explanation and authentication. Our ethical systems may be no more than cultural embellishments of a genetically programmed set of priorities. Pugh (1977) sees them as '*Sollwerte*', such as must be wired into any decision making system like the human brain. Our brains, then, are just like computers that have to take certain decisions, to do which they need priorities programmed in. The genes responsible for much of that programme make it responsible to them. Pugh, a computer scientist, sees no essential difference between the operations of the human brain and those of computers programmed to assist with complicated decision-making.

Pugh recognises as primary values like pain, fear, hunger, thirst, shame, sadness, bad tastes and smells, etc. On the positive side we have joy, pride, comfort, good tastes and smells and the sensations associated with the overcoming of fear, hunger, thirst, etc. Furthermore in strict analogy to the computer programmes for efficient decision-making, we employ also various secondary values. These arise, consciously or unconsciously, as adaptations to recurring situations in which it would be inefficient to 're-compute' one's decisions in accordance with the primary values. Secondary values enable quick and by and large judicious choices and actions. Collections of such secondary values can be found in every system of law, in religious texts and in folk wisdom. It is a moot point whether they are exclusively 'cultural', but we shall assume they are.

The secondary values are the ones we can tinker with. They are unlikely, however, ever to come into serious conflict with the primary ones. Primary values are innate and probably well-buffered. It would take great environmental manipulation or a long time of genetic selection to change them.

The feedback relationship between the secondary, cultural and the primary, innate values is very nicely described in a little book by H. R. Pulliam and C. H. Dunford (1980), an attempted synthesis of social psychology and sociobiology. If there is a big time-lag between a given cultural innovation or the establishment of new secondary values, and the felt effects of this innovation on the system of primary values, then cultural changes can lead to the demise of whole populations.

Perhaps what sets man apart from other social species is his ability to talk about his values with his fellows. Can his linguistic reflectiveness now enable him to alter the values inherited genetically through evolution? Dawkins (1976), like many other sociobiologists in the closing paragraphs of their cynical tracts, throws his reader the following crumb of comfort:

> We have the power to defy the selfish genes of our birth, and, if necessary, the selfish memes of our indoctrination. We can even discuss ways of deliberately cultivating and nurturing pure disinterested altruism – something that has no place in nature, something that has never existed before in the whole history of the world.

Ghiselin (1974) likewise concludes:

> If, as we have some reason to think, selection within the context of society has somewhat elevated our baser sentiments, and if, as the facts suggest, self-interest and common welfare are not fundamentally beyond reconciliation, we can reasonably hope to develop ethical standards consistent with biological reality.

But perhaps the reader can detect greater moderation in Ghiselin than in Dawkins. Consistency does not amount to transcendence or defiance. Certainly our values, once coaxed into verbal form, can have great potency in the personal and social consciousness. But according to Campbell (1975) they attain to this in the course of a process that has a sound evolutionary basis, albeit a cultural one. For Campbell, the language of uncompromised absolutes in the moral sphere evolved in order to counterbalance the egoistic tendencies which he thinks biological selection favours, but which threaten the stability of that complex urban life produced by cultural selection. Campbell, of course, ignores the possibility of a feedback between cultural and biological evolution. He sees the tension or opposition between biological inclination and cultural requirements as resulting in that admixture of self-interest and moral concern for others so easily discernible in social life. For Campbell, the causal role of language in the evolution of morality is collectivist. It has to do with conditioning and indoctrination, helping to compromise our biological tendencies. What is not clear, however, is whether it provides a logical grounding of moral categories, or whether the deliverances of reflective reason that it enables could ever allow man – individually or collectively – to 'transcend' his biology. Thinking of our values as grand products of culture controlling our baser nature may, of course, have survival value. But moral self-congratulation will never flatten the Gaussian hump of normal behaviour. The pure disinterested altruists among us – should any exist – form too insignificant a minority, despite the historical potency of their accumulated actions, to call for special confirmation in the evolutionary order. Like the necrophiliacs, the cannibals and the gifted idiots in our midst, they form a deviant residue. The genetic basis of their rare behaviour will itself be rare combinations of certain genes; and the circumstances that witnessed their rare acts will be evolutionarily rare.

A decisive test of Dawkins's hope that mankind can defy his selfish genes looms in what Hardin (1968) has called the Tragedy of the Commons. Resource exhaustion and population crash almost always occur whenever individuals have more to gain in the short term by selfish overconsumption than they stand to lose through the co-operative restraint required to conserve those resources. The tragedy of the commons is a Gordian knot that can only be cut by group selection or moral sanction culturally imposed.

But the group selective forces that threaten to be unleashed are awful to behold. If none of mankind is to suffer an agonising and untimely end, then purely moral persuasion of the Campbellian kind is the only possibility that remains. The future of the commons – spaceship earth, with its fisheries and wheat fields, minerals and oil – is finely, if not precariously, balanced. The first salvo in the battle by culture to avert genetic disaster was fired by Chief Seattle (1854) in his remarkable oration in reply to thee then President of the USA: near poetic proof of the altruistic wisdom of the biophile, and his keen insight into the ultimately self-destructive nature of entrepreneurial energies.

We do not regard sociobiology as having the last word on human nature. We side with its critics in agreeing that much of the speculation that has taken place has been somewhat wild. But in so far as one believes that one should carry on with research into such a complex problem as human nature, it is only fair to consider it not just on its own merits, but in comparison or contrast with its competitors. Anyone who has ever read Freud or Marx or Sartre, to name a few of the more influential theorists, at least one or two of whom were mildly concerned with the problem of empirical evidence, would perhaps agree that sociobiology does not fare too badly.

3 ETHICS

Kant (1785) in the *Groundwork of the Metaphysics of Morals* claims that reason could exercise only a 'weak and deceptive control' over the appetitive constitution – our desires, cravings, longings: that is, our basic biological needs. Nature would have

> prevented reason from creeping into practical employment, and from having the temerity to design on its own, with its weak insights, the model of happiness and of the means to achieve it; Nature would herself have taken over not only the choice of goals but also that of the means, and would with wise foresight have entrusted both solely to instinct.

At least one biologist has disagreed. In *Beyond the Mirror* Lorenz (1973) accounts for the gradual evolution of reason and consciousness from pre-existing instinctual components and behavioural routines. There is no great divide between instinct and reason, nor any battle between them for control. Both can contribute to wellbeing and happiness, to the choice of proper ends and means. Thus the evolutionary picture is one of interpenetrating overlays of sentience, appetite and reason. There is an ultimate phylogenetic strain of truth in Hume's celebrated claim that 'reason is the slave of the passions'. But the proximate causes of individual actions combine elements of both reason and appetite. We may note an analogy here with our account of biological and cultural evolution. Culture is more than a veneer; we are veritably steeped in it. Very little of human behaviour can be described as such without employing the terminology of the cultural. But culture ultimately is subject to the interests of the genes.

The biologist Stent (1978) continually chides his colleagues for blurring or overlooking the fact-value distinction. But sociobiologists talking about values or the evaluation of social acts do not have in mind the rather subjective and mentalistic notions of social scientists and moral philosophers. As Singer (1981) puts it: 'Sociobiology says nothing about motivation, for it remains on the level of the objective consequences of types of behaviour.' Stent clarifies the well-known distinction with an example of two boys, one of whom is seen giving the other a piece of sugar. As many writers besides Stent have observed, one would not praise the donor if one knew that he knew that the recipient was a diabetic. Moral appraisal depends on appropriate description. But so too does evolutionary appraisal of adaptive value.

Stent thinks that sociobiologists have no right to tread upon the domain of moral philosophy. Sociobiologists, however, concern themselves not with the morality of individual acts, but only with their motivation. Motivation can often be measured objectively. If it had no demonstrable behavioural consequences, its existence would be a miracle to the ethologist and psychologist. Both have long been measuring motivation in animals and man. With some increase in methodological sophistication, it might be possible to make room in experimental design for the motivational differences between the subjects that sociobiologists study. We believe, however, that this is not necessary, because there is probably a positive statistical correlation between the motivation underlying an act and the expression of that act. The more individuals behave like altruists, the higher their altruistic motivation must be.

All that matters for the sociobiologist when he considers the selfishness of an act are the selective conditions under which it arose. Side effects, so long as they have no selective consequences, do not enter the picture. Taking candy from a baby may prevent tooth decay; but it is highly unlikely that the prevention of tooth decay has had any evolutionary impact on the phylogeny of give and take.

Every instance of external altruism is due to kin-selection, group-selection, selection for reciprocity or pathology; in other words, it is not genuine disinterested altruism. What now remains of Stent's position is that individuals could still differ in their self-appraisal. Some would feel that they had been highly altruistic, that they had lost a lot in giving a piece of sugar to a friend. Others would not be moved unduly; they would hardly even register the act. We agree with Stent that this is what is important for the moralist, and quite elusive for the sociobiologist. But we do not agree that the sociobiological considerations we have just elaborated are of no importance to the moralist. It must surely interest him that all his important reflections on morality come from a mirror of culture that prevents light from falling on its objective ground. Moral appraisal surely rests on some external, objective pattern of acts. Why try to save genuine altruism from the sociobiological 'onslaught'? The subjective appraisal of altruistic acts must also be a result of evolutionary processes. Its function is probably only to afford individuals a constant check on their own degree of reciprocal altruism in comparison to that of others. No one likes being cheated by others. No one likes being deceived by others masquerading as altruists. On the other hand, it can do no harm to believe oneself to have been altruistic; for that way one can convince others of one's virtue (cf. Alexander, 1979, on self-deception.)

Recent responses by moral philosophers to the encroachment of ethology and sociobiology into the province of their own concerns has underlined some old themes and brought forth some new confusions. The confusions result first from misunderstanding the claims and methods of those disciplines, and secondly from fear that their intellectual territory will be invaded. One philosopher, Midgley (1979), has launched blistering and condescending attacks on sociobiologists. She has taken Dawkins to task for his 'personification of little bits of goo'. In a diatribe against Wilson (Midgley 1980) she advances no arguments against the substantive theorising at issue. She lets most of her invective turn on some of his inappropriate choices of words on unimportant points. Her diatribe endes with the overpowering insight that 'elephants cannot be caught in jam-jars'.

Other philosophers have been a little less excited but still interestingly misguided in their responses to the new discipline. Mackie (1981) and McGinn (1979), for example, reported themselves as impressed by Dawkins's book *The Selfish Gene*. But McGinn so took its message to heart that he thought it was 'easy to see why a gene for group directed altruism would not stay around for long', while at the same time dogmatically asserting the existence of 'pure, disinterested altruism' in human beings and his faith in the 'power of genuinely moral ideals in human thought and action'. He insists, moreover, on the importance of the cognitive aspect of morality, thinking it a by-product of evolutionarily useful intelligence. He does so after peculiarly attributing to group selectionists the view that group selection has produced or would produce a morality based on a wholly appetitive foundation. Mackie's mistake is to think that

> the issue is this: is there natural selection by and for group survival or species survival *as opposed to* selection by and for individual survival, (or, as we shall see, gene survival)? Is behaviour that helps the group or the species, *rather than* the individual animal, rewarded by natural selection which determines the course of evolution? (our emphases)

To these confusions and misunderstandings on the part of two respected philosophers can be added Midgely's claim (1979), made in apparent ignorance of Trivers's work, that the selfish gene account is powerless to explain the apparently one way, unreciprocated tender loving care that parents heap on their offspring.

The conceptual issues need some untangling here. We shall in due course have something to say about the fact-value distinction. We have already shown sociobiological considerations to be legitimate in complementing proximate accounts of human behaviour with ultimate ones. We have also cleared up misunderstandings of the selection mechanism governing behavioural evolution, and entered some moderating considerations about the semantics of altruism in everyday and biological reasoning. We have explained how group selection may be a force to be reckoned with and need not always act in opposition to the forces of gene selection. We have discussed the existence of pure disinterested altruism and given good reasons why such freakishness would have no phylogeny. Finally, given the phylogenetic importance of the passionate priorities which reason must subserve, it is unlikely that purely ratiocinative intelligence would have evolved without intimate connections with our motivational states. Morality, construed as the rich deliverance of Kantian reason (*Vernunft*), is, if an evolutionary product at all, not simply a useful by-product but rather a central adaptation of our species.

Wickler (1971) showed that those of the Ten Commandments that concern social behaviour (as opposed to behaviour towards God) appear to inform the social life of many other animal species. Thus our social norms probably have a biological basis. If now freed therefrom, this could only be possible because of their continuing conformity to biological needs; but one might then also wonder why they ever needed to be freed from the biological substrate at all. Most probably biological tendencies towards certain forms of behaviour and social arrangements became solidified and sometimes modified within culturally imposed norms; with the latter ever subject, however, to continuing control from below.

Against McGinn and Mackie, we maintain that all four following possibilities appear to be open:

1 Group selection alone might have produced morality with a wholly appetitive basis.
2 Gene selection likewise.
3 Group selection might have produced morality with a wholly cognitive basis.

4 Gene selection likewise.

We prefer, however, a fifth possibility, an obvious and catholic one. Both gene selection and group selection might have been at work, producing a morality based on appetitive and cognitive faculties inextricably spliced after eons of evolution. Humeans and Kantians cannot bend biology to decide between their analyses of morality. It is difficult, if not impossible, to conceive of a purely cognitive creature possessed of a will, for which one could make Kant's distinction between merely '*pflichtmässigen*' actions (actions apparently in accordance with what duty requires) and actions '*aus Pflicht*' (actions performed for the sake of duty). Even when McGinn describes such a creature in a thought experiment, he endows it at least with the desires for food and sex. But who knows but that every desire in the Shakespearian corpus may be wrought by the alchemy of means–end analysis working on those two powerful ingredients!

The rational constraints on belief and perception, and the power to make inferences to likely consequences of one's actions, are imperative for the efficient satisfaction of desires. Our cognitive faculties range from perception heavily aided by inference, to inference heavily aided by perception. And desires obtrude at every point. We often see what we want to see, and believe what we wish to be the case. Intellectual prejudices might often be regarded as the result of high-level desires. As Mill (1863) pointed out, our desires can be deflected from ends to means, so that money and knowledge can be desired as ends in themselves. Metaphors of sense perception and desire permeate even our highest intellectual processes. We grasp the insights of relativity theory, get the feel of its formalism, and derive enormous satisfaction from doing so. The infant wants everything it sees, the wise man sees everything he wants. The intervening route goes up the slope of reason.

Our brains might extend Ghiselian gonads; but also as one evolutionist puts it: 'The ape who could not see the branch to grasp it was very quickly a dead ape.' We gave an analysis in the first chapter of what it means to say that a selective force of such and such a kind is at work. We must be wary of any uniqueness claim for sway over all species. But we doubt that resolving the neutralism-selectionism controversy would be at all relevant to our theorising about human morality.

How then does biological theory bear on ethics? We have argued that it is not especially helpful to see recent development in sociobiology as raising once more the historical opposition between man's passions and his reason, by mistakenly linking each of these accounts with a different stance in the selectionist controversy. A much more fruitful distinction, to our minds, and one with respect to which those two great philosophers can still usefully be contrasted, is the distinction between evolutionary ethics and the evolution of ethics. A system of *evolutionary ethics* is one based on a criterion of value purportedly derived from the evolutionary theory of the origin and proliferation of life on earth. It was in this connection that social Darwinists became notorious. In putting forward theories of this type, they committed the naturalistic fallacy in a form so crude that even modern doubts as to its general fallaciousness would afford no defence. By contrast sociobiology is the evolutionary theory of the origin and stability of social behaviour. When fully developed, it may account for the *evolution of ethics*. It certainly does not purport to derive any ethical code from the theory of our evolution. Thus it is a simple conceptual mistake to think of sociobiology as inherently racist, sexist, or elitist. Scientific knowledge has always been perverted for political ends. The proper antidote to neo-fascist abuse of this new theory of human nature ought surely to be a fuller understanding of its scope and limits. The very same knowledge could after all be heeded when planning social and political reform for the greater happiness of all.

Evolutionary ethics

Social Darwinists fallaciously derived behavioural prescriptions from very questionable premises, from distorted slogans summarising the early theory of evolution. That life was a competition in which the fittest survived by dint of superior strength, cunning and aggression was somehow supposed to justify the laws of the socio-economic jungle. But cannot some systems of evolutionary ethics do a little better? Can we not discern in the evolutionary process, knowing what we do about the underlying selective mechanisms, some criterion of value on which an ethical code might be based?

Quinton (1965) reached the measured conclusion that we cannot. According to him, evolutionary theory affords at most 'technical' values, which can, in certain cases, conflict with one another. They must therefore be subject to some final value. He maintains that in the writings of Huxley, Waddington, and similar theorists there is always an implicit appeal in the final analysis to some notion of value of this kind, assumed to exist but transcending all biological standards. The inadequate technical values that always need such supplementation all have an implicit appetitive element and are anyway vague and ambiguous. Values such as the survival of the species or the progressive complexity of living forms deliver themselves of deductive endorsement of ethical laws only with the greatest difficulty, if at all. More recently, Flew's *Evolutionary Ethics* (1967) arrives at similar conclusions.

Quinton mistakenly maintains that 'the evolving and naturally selected human characteristics which are principally relevant to ethics and social policy are socially acquired and not genetically inherited'. This statement is obscure and unsupported. It now appears fragile and dated before the recent wave of theorising in sociobiology. As shown above, the powerful theoretical tools of kin-selection, reciprocation theory, stable strategies and the like also reveal how wrong Flew was when claiming that biology would yield only 'mild vegetarian' notions for the study of ethics.

Can we say with hindsight that evolutionary ethics is an intellectual dodo? Modern biology might well be able to contribute something to the design – if not the logical grounding – of ethical systems. The Socratic injunction to know oneself, in one's personal idiosyncrasies and by virtue of being the kind of creature that one is; the Aristotelian emphasis on the moral discretion that derives from professional skill and experience; utilitarians' emphasis on happiness, or the satisfaction of desires, and Mill's worry especially in identifying which desires were meet and right to fulfil; act utilitarians' emphasis on the actual consequence of one's actions, and the consequent importance of foresight and understanding of the causal web; rule utilitarians' emphasis on what in effect amount to optimising behavioural strategies; the dangers inherent in relativists' regarding bizarre and barbaric cultural practices as a moral *fait accompli*; the emotivists' inability to summon up the right sobs and laughs to steer the destiny of mankind; the egoists' failure to realise that evolution holds some aces; the sceptics' refusal to grant the existence of any worthwhile purposes or stable means of achieving them; the analytical philosophers' tinkering with the linguistic surface, and disinclination to delve more deeply into the biological origins of human values; all are cast in a new light when one tries to look at the human predicament through the several lenses ground on recently broken ground in biology. With the new knowledge at hand, Kant's system based on the categorical imperative provides a distinct possibility of resuscitating a form of evolutionary ethics.

Kant proposed his categorical imperative as the means of establishing the moral worth of the maxims by which we conduct ourselves: 'Always act in such a way as though you could will your maxim to be a general law'. But in certain applications it stands in need of supplementation. As Korner (1955) has observed, we should not ask 'whether the universalised maxim by itself is contradictory, but whether it *in conjunction with other true statements about the world* is logically possible' (p. 148). He goes on immediately to remark that 'The difficulty of this interpretation lies in the question which statements about the world are to be considered in order to judge of the morality of an action.' An obvious candidate is the body of biological knowledge about man not just as a physical organism but also as a social animal. Population genetics is more relevant to the incest taboo than any knowledge in Kant's day. Modern knowledge about lethal recessives and the consequences of close inbreeding provides an immediate explanation of the origin *and value* of the incest taboo. We need only have some criterion of value that allows that the world is a better place without unhappy individuals in order to convert the explanation into a justification. (The mutants in question are homozygous for disadvantageous recessive mutations.)

Pugh (1977) sees the non-linearity of biological utility functions as an underlying cause of the social ideal of distributive justice. This is the implicit message in advertising aimed at the affluent West by international aid organisations. A small monetary donation from a Westerner will buy provisions that can mean the difference between life and death for a pauper elsewhere. Within any one society distributive justice may be feasible if it is producing a surplus. But Hardin (1977) has argued that in 'desperate' communities, to avoid a population crash, the guardians of resources must be better fed and provided for in order to act 'in the interests of posterity'. He goes so far as to claim that 'distributional justice is a luxury that cannot be afforded by a country in which population overwhelms the resource base'.

A proper assessment of the ontogenetic stability, individual variation and systematic ramifications of socially important traits is a precondition for successful policy-making. On present evidence the spectrum of opinion seems to be as wide as ever. On the one hand Skinner (1971) advocates environmental engineering as the way to create happiness in infinitely plastic natures. On the other, Darlington (1978) vituperates against comprehensivists in schooling for not taking into account genetic variation in educability of our children – among them apparently, some very busy arsonists somehow dimly apprised of the same facts. As someone has said, however, the only answer to dangerous knowledge is more knowledge. Even if our theories today indicate no definite policy directions, at least we should be beginning to appreciate the importance of a particular kind of consideration. This is that our basic behavioural biology now finds itself in newly created niches in which it may be ill adapted. The hope remains that we shall be able to chart the pitfalls and avalanches on our epigenetic landscape that threaten those acting in accordance with the wrong universalised maxims. This, indeed, is what determines those maxims as wrong. When Kant enjoins us to ask whether the maxim by which we are acting could consistently be willed to become a universal maxim, the sorts of possibilities considered when judging thus of consistency must be the physical possibilities admitted by the laws and facts of nature. Among these are the laws of genetics, and the fact of evolution past and future. If policymakers were to become more aware of these laws and facts, then evolutionary theory would finally become a source of grounded wisdom about our own nature. Our genes not only determine, but also determine our emotional responses to, what is alien to our nature.

As Campbell (1975) has observed, no ethical theory has ever managed to provide logical grounds for the values that we have – let alone the values we ought to have. Evolutionary theory, in the role envisaged above, is not being pressed into service as the sole premise from which values are being derived, thereby crudely committing the naturalistic fallacy. All it does is make us more mindful of the *given* in human nature, the reasons why it *is* given, and the limits of the human condition as such. Ultimately transcendent values are still in the background when applying the categorical imperative. For example, the preservation of our own species might not be of any earthly biological value at all; and it is hardly a value served up by evolutionary theory as such. (Indeed, it was precisely when Teilhard saw man as the perfection of the evolutionary process that he most clearly ceased to be scientific.) Thus any application of Kant's categorical imperative with the due regard advocated for the body of biological knowledge now at our disposal, provided that it is further premised upon our continued survival as an overriding and additional value, does not in any way commit the naturalistic fallacy.

Another fallacy to which some thinkers are prone when considering the links between biology and ethics is what might be called the *anti-naturalistic* fallacy. Quinton (1965, p. 128) provides a good example:

> by treating morality and moral agency as natural phenomena, by considering morality as a functioning social institution, *and so firmly*

> *connecting it to our desires and satisfactions* (this approach) repudiates the essentially other-worldly conception of morality which is characteristic of intuitionism. (our emphasis)

There is no reason at all for thus downgrading, from the proximate standpoint (or indeed the evolutionary one) the importance of the cognitive element, as was argued above. 'Does nature . . . create a rational creature without entrusting anything to the operation of his reason?' (Hume, 1977, p. 161). Hume's question has more point in the present context if we replace 'anything' with 'any part of his morality'.

We need not think that the function of reason in moral matters, a function conferred on it by evolution, has always to be that of excogitating the consequences of one's actions in an emotionally detached way. Much of our moral sense, or moral knowledge, need not be of an articulable, propositional kind at all. Pure reflection might yield or uncover synthetic *a priori* truths in the moral domain. But the fundamental categories of moral thought and forms of moral intuition could be part of our innate genetic endowment, in the way Lorenz (1973) suggests is the case with the categories of substance and causality, and with the forms of physical intuition – space and time. Morality might be thus innate because genetically based, and crucial for our past survival in an increasingly complex social order. As Quinton has observed, 'evolutionary considerations do not rule out a moral sense theory which conceives the moral sense in rather close analogy to the physical senses'.

Nevertheless it need not be, as according to McGinn out of Kant, that, once equipped with reason and intelligence, we must acknowledge the moral claims of other rational beings. We might simply be genetically rigged and culturally conditioned so to acknowledge. There need be no compelling transition, such as is urged by Sprigge (1979), between a proper cognitive grasp of the reality of other persons and having the wish expressed by the imperative to treat other persons as ends. The genes responsible for the former might be responsible for the latter, in which case the grasp and the wish would have no internal logical relation to one another. And as one might expect from man's symbiotic evolution with domestic animals, there could be polygenes resulting in a small fringe calling themselves animal liberationists.

The evolution of ethics

Thus far we have confined our discussion to evolutionary ethics. It is now time to say something in general about the evolution *of* ethics. Sociobiology is a good example of an evolutionary theory of the origin and stability of social behaviour in all species. It does *not* offer any criteria of value or normative laws. Its laws are descriptive; and, in some cases, cynically re-descriptive, especially in the case of altruism.

But is not altruism altruism, under any causal account? We need, by way of concession to the scientist, a phenomenologically correct account of group life using our everyday vocabulary of 'altruism', 'selfishness', 'spite', etc. Even these first order descriptions of behaviour will be hard to agree upon. There are all the problems of translation in the jungle cases, of self-deception and rationalisation and intentionality and so on. But once we have such a description of a stretch of group behaviour, and have distilled important dispositions, we must ask whether any causal explanation of those dispositions must, when couched in evolutionary terms, save the appearances. Will mother love have to be regarded as any the less loving for our having a compelling story of its evolution via 'selfish' genes? This is no more than the well-known problem of theory-dependent description. A good example is that of warning cries among birds. There are as many different theories of its evolution as there are conflicting descriptions of it as 'altruistic', 'selfishly opportunistic', and so on.

In so far as our ethical codes are manifest in our behaviour towards our fellows, sociobiologists are entitled to call certain dispositions moral, and to frame evolutionary explanations of their origins and stability. They thus postulate genetical bases for various kinds of altruism and account for their prevalence and constancy in our species in terms of the selective advantage they confer. Moreover, as we have seen, their game theoretical methods throw fascinating light on why mixtures of the gentle and the caddish should be evolutionarily stable. It matters not that so far only very simple 'one gene' or 'two gene' bases for these behavioural dispositions have been considered (*pace* Midgley).

We have seen in chapter I that there is a genetical basis in the first sense to every disposition and in the second sense to virtually every behavioural disposition. It does not matter how many genes are involved, nor how complicated their interactions might be. If a behavioural trait displays heritability and variability in a population of differentially reproducing individuals then it will respond in its distribution to selective pressure.

It is the whole spectrum of our social behaviour from which we are to extract that system or systems of '*Sollwerte*' constitutive of our humanity biologically understood. We must not confine our attention only to that part of linguistic behaviour to do with the formation of moral codes of whatever degree of other-worldliness. For the purposes of an empirical study of morals we must attend not only to the first order transactions eligible for ethical labels, but also to the higher order ones. These range from the simplest statements we make to children about moral matters, through articulations of codes, to conceptual analysis and the search for fundamental categories and axioms in moral theory. All these are unique to man by virtue of his language.

There is an obvious plurality of observable *patterns* and of normative *codes* of behaviour. Whether at root there is a common core or master plan to be discovered, we know not. There is an analogy here with observable patterns of speech and of normative codes (grammars of competence) in linguistic theory. What people actually utter, with ums and ahs and other infelicities, does not detract from their guiding conception of a grammatically correct sentence, even though this be a controversial notion. Similarly, how people deal with one another daily contrasts with their guiding conception – no doubt strongly determined by culture – of what constitutes virtuous behaviour.

By way of concession to the relativist, we can expect the biological study of human social behaviour to issue in a classification of patterns and codes that are genuinely distinct, even though, as ethologists have made us aware, there are common denominators of gesture and so on that one can trace back to the apes. The relativist position is even more plausible if we are prepared to see behavioural consequences in the likely genetic diversity between the human populations of different cultures. The non-existence of moral universals, of course, would *not* show morality to be biologically undetermined. For genetic variation could well underlie the variation in morality. Thus Baker (1974, p. 426):

> One wants to know whether it is conceivable that members of two taxa may differ in large numbers of groups of genes affecting many different parts of the body, but not at all in those that affect the nervous and sensory systems and therefore play a part in determining mental qualities.

Once one accepts that there must be a physiological basis for our dispositions to social behaviour, one may ask Baker's question with respect to cross-cultural variations in *morality*.

Moreover, if one thinks that we can and ought to follow Dawkins's wan rallying call to Homo sapiens to defy his genes, there is no reason why we shouldn't take up anti-relativist cudgels and make members of other cultures give up certain of their practices, no matter how much this might go against their natures. And the same holds for different groups within one and the same culture too.

Hume's own advocation of the sort of empirical programme sketched above, despite his emphasis on the appetitive and his faith in the existence of a set of universal principles, is ripe for contemporary vindication in the 'New Synthesis'. In the literature of anthropology, psychology and sociology there must be a battery of behavioural grammars calling for comparison and classification. This is the challenging research project urged by Campbell in the paper cited above. Even if it fails to provide a set of universal principles, it would nevertheless be fair to find in Hume that cast of mind that can produce the natural explanation. For these behavioural grammars represent a group's adaptive solution to the challenges that have shaped its social life. And Hume (1977, p. 188) himself was crucially aware of the context-dependence of these grammars:

> Thus, the rules of equity or justice depend entirely on the particular state and condition in which men are placed, and owe their origin and existence to that utility, which results to the public from their strict and regular observance. Reverse, in any considerable circumstance, the condition of men; Produce extreme abundance or extreme necessity: Implant in the human breast perfect moderation and humanity, or perfect rapaciousness and malice: By rendering justice totally *useless*, you thereby totally destroy its essence, and suspend its obligation on mankind.

Is this the sad story of the Ik of Uganda? (Their desperate physical and moral plight, however, has yet to be confirmed by independent observers after Turnbull's initial report (1972), which was rather vague and partially contradictory.) One cannot be sure of their social history in earlier, more hospitable environments. But if, as Alexander (1979, p. 64) conjectures, social structure is primarily the evolutionary response to predation and only secondarily to the exigencies of resource distribution, one might on theoretical grounds regard the dissolution of morals among the Ik (if they ever had any) as quite unsurprising. 'Under severe survival conditions, morality disappears' (Hardin, 1977, p. 79). Present research into evolutionarily stable strategies will probably lend further support to Hume's claim that 'The common situation of society is a medium among all these extremes.' We contend that the Humean approach may thus find its full development in sociobiology, the theory of the evolution *of* ethics. And we argued above that its results may then be pressed into badly needed service in a Kantian form of evolutionary ethics. There is no reason at all why the new science of morals should not become part of the moral sciences.

III

EVOLUTIONARY EPISTEMOLOGY

1 INTRODUCTION

Antidotes to skepticism can strike one as silly.

> we stood talking for some time together of Bishop Berkeley's ingenious sophistry to prove the non-existence of matter, and that everything in the universe is merely ideal. I observed, that though we are satisfied his doctrine is not true, it is impossible to refute it. I never shall forget the alacrity with which Johnson answered, striking his foot with mighty force against a large stone, till he rebounded from it, 'I refute it *thus*.' (Boswell, 1953 edition, p. 333)

G. E. Moore produced his own famous response to skepticism in a lecture to the British Academy. He held up his hands. Not, though, to surrender to the forces of skepticism, but to

> '... prove now, ... that two human hands exist... [by] holding up my two hands, and saying, as I make a certain gesture with the right hand, "Here is one hand", and adding, as I make a certain gesture with the left, "and here is another".' (1959, pp. 145-6)

Just as Monsieur Jourdain discovered that he had been speaking prose all his life, so the man in the street is told now that he is daily engaged in scientific theorising, with sensible concepts tailored to his needs and to the world around him. The untutored response to the skeptic is to leave him to his folly. This response is by and large developed by Wittgenstein in *On Certainty*. The pendula of doubt need their lynch-pins, the seesaws of uncertainty their fulcra. Let the skeptic live out his convictions, or absence of conviction. Let him walk into walls, or into the path of an oncoming car. In brief, let his theory of knowledge be his known theory, his guide to living. If he continues to behave in ways 'rational', ways that enable him to satisfy his desires and achieve his goals, he is after all one of us. This challenge pierces the prison sphere of solipsism, and it collapses to the plane of public life.

Evolution may be looked upon as a vast catalogue of Johnsonian encounters with stones. It has taught us appropriate intellectual responses to various contacts and collisions, by structuring the brain of the mind that responds. This is the central claim of evolutionary epistemology. Its modern revival began with Lorenz's paper (1941) on the phylogenetically *a posteriori* nature of the Kantian *a prioris*. Apart from the work of Campbell (1966), developing ideas of Popper (1962) into what amounts to something like memic evolution by natural selection as discussed in chapter II, the main work in this area has been carried on by German thinkers. Lorenz (1973) gave a fuller presentation of his ideas in *Beyond the Mirror*. Vollmer (1975) gave a scientifically informed and philosophically sensitive account of the subject in his as yet untranslated book *Evolutionäre Erkenntnistheorie*. Riedl (1980) stole the popular market with his *Biologie der Erkenntnis*, soon to appear in English. A work roughly in the same tradition as far as choice of problems, but not methods or terminology is concerned, was produced independently by the American anthropologists Laughlin and D'Aquili, entitled *Biogenetic Structuralism* (1974).

The main claims of evolutionary epistemology may be summarised as follows:

1 Every successful advance in adaptation corresponds to an increase of information about the domain for which it is significant.
2 Evolution is a knowledge or information increasing process.
3 Cognitive structures, like any other biological organs, evolve as a result of natural selection. Consequently one expects
   (a) that these structures will be tolerably adapted to the 'reality' that is of importance for the species concerned;
   (b) more recent structures must have brought a selective advantage;
   (c) certain items of knowledge or forms of knowledge gain can be inborn.

4 Although inborn human knowedge is in the most immediate sense possessed by the brain (as the seat of mentality) it is also, in the more recondite sense mentioned above, somehow contained in the DNA of a cell nucleus.
5 Cognitive structures, both conscious and unconscious, are accessible to scientific investigation.
6 Our conscious cognitive faculties are the most recent overlay; they have perhaps been subjected to the least selection. Because of the increase of what can be grasped and reflected upon, much of its behaviour will be facultative, and not necessarily adaptive; and yet there may still be fundamental limitations on what we can cognise, environmental boundaries to pure reason.

Most scientific advances are accompanied by conceptual innovations that subsume into a more pervasive category separate manifestions that had been thought disparate. We need only think of the scientific notion of energy, and the disparate forms it can take – motion, heat, coiled springs, chemical bonding and so on. It is a weak objection to such conceptual unification that that is not what 'human energy' is like. The innovation analogous to this in the case of evolutionary epistemology has not been made very explicit by the practitioners themselves, but is central to their enterprise. This is to equate increase of knowledge (or of information) at the molecular level with that at the neurological level, and also presumably with that at the level of linguistic expression. Riedl (1980), for example, speaks of information at the molecular level being learned, by being anchored in the genome. He envisages a continuum of some recondite quantity ranging in expression from genetic information to human knowledge. It may transpire that the contribution of evolutionary epistemology turns out to be precisely this unification of the concept of information, and its application at all biological levels. But the information we are concerned with here is not that of the so-called information sciences that are concerned with the ratio of information to noise in the transmission of signals within some medium of communication. For the latter notion is precisely measurable (in units called 'bits') and has little to do with the semantic novelty or value that a piece of information ordinarily so called can have for a person in a certain situation, already holding certain beliefs. Similarly in the genetic context the information which is on the evolutionary increase is of value only because of certain selection conditions that prevail. It is useless in the absence of an energy flow into the organism, securing its morphological expression. Much of it would be useless also were the organism to be placed in a radically different environment in which all its closed behavioural programmes backfired on it. From the lowest organisms to the highest, coded or learned information counts as valuable gain only because it renders outside happenings less strange and unexpected. It is therefore strongly dependent on environmental conditions.

Lorenz (1973) has given a persuasive account of cognitive mechanisms that have been selected for particular domains, and has addressed himself to the question as to how those mechanisms developed that are the functional preconditions of our own reason. Every cognitive structure has a genetic base. Even the most abstract process of learning requires neuronal structures to make it at all possible. The empiricist conception of the mind as a 'tabula rasa', a clean slate, is simply untenable. A long and diverse philosophical tradition attests to the force of the opposing view: Plato's theory of forms, Aristotle's syllogistic, Bacon's *idola tribus*, Humean instincts, the rationalists' taxonomies and necessary truths, Piagetian cognitive structures, Jungian archetypes, Lévi-Strauss's cultural universals, and Chomsky's rule schemata in grammar. We are not interested here in whether it is exegetically accurate to say that their inventors regarded them as innate in the appropriate biological sense canvassed here; for certainly each has been held at one time or another to be so.

We must be mindful once again of our earlier distinction between two senses of *heritable*, namely the first, ontogenetic sense and the second, quantitative-genetic sense. If an empiricist like Locke imagines that all cognitive differences between human beings are attributable to environmental differences, the matter can only be put to empirical test. This is the residual interest still held by the proposition that our cognitive capacities are 'innate'. For it hardly merits argument that the 'slate' must be endowed with requisite structure from the very beginning of ontogenesis, in order to be able to receive and process information at all.

Moreover, this slate is probably the product of step-wise evolution. This view is supported by the elegant experiments of comparative psychologists on species-specific learning (Bitterman, 1965). These show how different mammals learned different tasks at different rates. A very simple example one could give to illustrate the general point at issue here is as follows. It is very easy to teach a pig to dig for truffles.

A dog can also manage. But a seal has the greatest difficulties, not because of its relative immobility on land or for want of a keen sense of smell, but because it has never had to dig and scrape for food, or to bury it for later unearthing and consumption. (We, of course, set aside the question of motivation here.)

Evolutionary epistemologists have drawn on the findings of comparative psychologists and ethologists, and have organised the behaviour patterns of many different species into a roughly phylogenetic lineage. There is a crude continuum of behavioural patterns or 'programmes' of varying complexity and flexibility (or 'openness'), which Lorenz (1973) has used to bring human mentality within an evolutionary line of fire. These include:

1 Kinesis
2 Phobis and Taxis
3 Imprinting
4 Sensitisation
5 Habituation
6 Trauma
7 Instinct
8 Unconditional reflex
9 Conditional reflex
10 Association
11 Imitation
12 Insightful learning
13 Proto-conscious behaviour
14 Conscious human action

This ordering is reflected roughly also in the phylogenetic order culminating in man. It is well known how rigid are the behavioural programmes of the social insects such as bees, ants and termites. By contrast the social behaviour of chimpanzees, although not necessarily any more complex, is certainly far more flexible. In general, one can discern in the evolution of species a trend from instinctual and inflexible programmes towards those of greater flexibility, less dependent on drives and more attuned to a wider range of environmental input. But this is not at all to say that man himself has been evolutionarily purified of instinctual behaviour, nor is it to play down the tremendous importance of the latter in his psychological make-up.

2 REDUCTIONISM AND SCIENTIFIC THEORIES

Scientific theories variously investigate physical objects or systems of objects ranging from fundamental particles to supergalaxies. Some of these might be regarded as more theoretical, such as electrons and atoms, others more observable, such as elephants and cabbages; this distinction will not concern us here. The objects may be ordered in various ways – by size, inclusion or containment, time of origin or genesis, and so on. At various 'levels' different sciences concern themselves with the laws or regularities both of the behaviour and of the structures of the entities concerned. Quantum physics investigates the behaviour of fundamental particles in different fields of force, and the structure of atoms. Molecular biology investigates the information-theoretic properties of large molecules, their rates of replication, mutation and so on. Cell biology investigates the chemical and thermodynamic properties of cells, their division and growth. Then we have the sciences of morphology, embryology and so on up to the relatively less 'exact' sciences of ethology and ecology. And from there on our scientific interests proliferate in a formidable array of aspectual glimpses afforded by the human sciences – cognitive psychology, organisational psychology, sociology, decision theory, and theoretical linguistics, to name but a few.

Each of these sciences has a characteristic subject matter – a choice of objects or systems – whose patterns of behaviour it seeks to understand by correctly classifying, noting regularities (especially under experimentally controlled conditions), and identifying functional, statistical or causal correlations and dependencies. Very often the key to behavioural regularity is internal structure; and, as evolutionary theory maintains, the key to internal structure is past behavioural regularities.

After ontology, we come to the second feature of scientific theories: a vocabulary of concepts and relations that enable us so to classify the objects above, to note regularities in their behaviour, and draw correlations calling for explanation. These concepts, of course, vary from theory to theory, not only across the ontological 'levels' distinguished above, but also perhaps across competing theories dealing with the same level.

The unity of science, however, lies not so much in the prospect of reduction – i.e., whether some preferred stock of concepts in a given theory can be used to devise definitions of all the concepts of any other science, and the latter's laws thereby derived from the former theory. It lies rather in a unity of method. In every science one encounters the same logical structure in the activities of hypothesis formation, experimental design, observational testing, statistical sampling, and theoretical adjustment in the light of new data. The canons of scientific method do not discriminate between large or small, visible or invisible. This is not at all impugned by the special observational difficulties in the subatomic realm, if one espouses a causal theory of perception anyway; indeterminacy in the small is just active observership writ large. Whatever view one might hold of the interconnections between prediction, explanation, description and understanding, we maintain it will be broadly the same across all levels of scientific subject matter.

The final feature to be emphasised about scientific theories is the kind of laws they contain. The Newtonian paradigm in the physics of charged mass points moving in gravitational or electromagnetic fields, with its emphasis on analytical methods for the solution of differential equations, has engendered an abiding misconception about the nature of laws of nature, and their role in explanation and prediction. *Science is not necessarily deterministic*. Modern scientific theories at all levels contain essentially probabilistic or statistical laws. The implicational potency of the single case is thereby severely reduced. We approach truth – or, in Popperian fashion, unearth falsity – by the method of sampling and of repeated trials. Quite what can be salvaged of the naive notion of cause (as a relation between events) in the probabilistic sciences is not clear.

The ordinary language paradigms of ignitions causing explosions, and of collisions causing reboundings do not emerge clearly from the regimented syntax of scientific theories. As Davidson (1967) has claimed, single causal links are pieces in a mosaic of nomic explanation. Event A caused event B just in case there are descriptions A′ B′ in the vocabulary of an appropriate physical theory according to which B′ is predictable on the basis of laws in that theory when A′ is assumed as given. It may well be that no one theory knits together the ravelled sleeve of cause; and that, *pace* the reductionist, one will need different levels of stitching to fashion a proper fit with the body of experience. But despite the probabilistic character of the causal web, there is a widespread belief (sometimes going hand in hand with reductionism) that all scientific theories should deliver causal narratives of happenings in their particular neighbourhoods or domains of application. Time's arrow should etch the network of causal dependencies, piercing any event of any moment. This picture may indeed be adequate for a common sense creature transacting with medium sized objects in a relatively stable nook of space time. It may indeed be just the scheme of ordering and understanding with which evolution had equipped us.

We do not wish here to dissect the anatomy of cause too far. For our present purposes it is enough to highlight just one overall feature of so-called *causal* explanation that appears to legitimate the demand that all scientific theories be causal theories. This feature is the following. In order that an explanation of an event (or state, or set of circumstances) B be a causal explanation, it is necessary and sufficient to cite an antecedent A accepted as true or probable by one's audience, and then to show that, according to the laws of the explaining theory, B would (probably) have arisen or occurred in due course: that is, later in time.

This rubric covers a large number of cases – ignitions causing explosions, collisions causing reboundings and so on, right up to the case of untimely deaths and accidental mutations *causing* adaptations of descendants. This last example is the most important one to cover with the analysis. For it was supposedly Darwin's great triumph to have given a *causal* explanation of teleological behaviour and adaptive physiology. Here the antecedent A to be cited is a manifold accretion of events of great and little moment, collectively bringing in their wake – according to the theory – the state or circumstance B of adaptedness. Parodying Kant, one might say that according to Darwin adaptation without evolution is empty; evolution, despite adaptation, is blind. If Darwin's is to be a paradigm of causal explanation, though, the A-B dipole of causal narrative is no shortest thread between datable knots. It is a huge chunk of the net itself, with 'adaptedness' somehow residing along the frontier of time's shadow as it slowly draws across. The set of circumstances that we call adaptedness is not any point event. Adaptedness reveals itself in a host of present actions and reactions of organisms in their environments. Adaptation itself involves shifts within sizeable pieces of the causal web.

What makes a discipline scientific is the ideal of objectivity, of publicity of postulate, repeatability of experiments, simplicity of structure, consistency with data, and – subserved by all these – falsifiability in principle. It is not so much the prospect of reducing the discipline in question to one more basic, such as physics. Most notably, the latter view has been opposed by social scientists in the hermeneutical tradition – the 'Verstehen' theorists or *Geisteswissenschaftler*. Their position is roughly as follows:

> Understanding human behaviour is a matter of apprehending various actions, from within a closed circle of intensional notions such as belief, desire and meaning. These notions supposedly have their own, special 'life', their own resonances of meaning, and all fit together or inter-animate to provide a view of matters denied to any other,

> 'external' scientific viewpoint. We must, for example, be able to understand people acting on false beliefs, or on the basis of the very theories we are trying to devise in explanation of their actions. We must always distinguish acts from mere physical happenings. The man making marks on a piece of paper may be drawing up his will; yet no amount of strictly physical description and theory will draw from the paper anything of a piece with the will drawn up.

This insistence on social facts, however, does not contradict our account of the unity of science, unless sound reasons be given why intensional explanations of the kind emphasised be not subject to the same logical demands as other scientific theorising. It appears to be no more than an anti-reductionist sentiment specifically concerning the level of human behaviour. Similar anti-reductionist claims might well be made by a cell biologist insisting on the indispensability of the idiom of coding and information exchange when accounting for cellular happenings. 'Seeing things under a particular aspect', using concepts that have a 'special life of their own', peculiar to a particular level, is not the prerogative of the social scientist. (For other arguments for the methodological congruence claimed above, see Hempel (1962) on the function of general laws in historical explanation, Stegmüller (1977) on the deductive similarities between poetic interpretation and cosmologicl theorising; and Føllesdal (1979).)

Finally, Hellman and Thompson (1975) have established the logical compatibility, at each transition between 'levels' as described above, of the claim that there is nothing but the physical ('matter in motion') and that all facts are determined by the physical facts, with the claim that concepts and laws at a higher level cannot be reduced to those of the lower level. They speak thus of ontological determination (that the physical exhausts all there is) and explanatory anti-reductionism (that concepts for the higher levels do indeed have a 'life of their own', and not just in the case of the human sciences). But even if the higher level sciences did not have niches specially carved for them in this logical fashion, their existence would be secured by the fundamental utility and good fit of their own special concepts. Our world picture is an heuristic hegemony, perhaps even because of human biological necessity. For selective advantage no doubt accrues as much to genes that fashion brains that understand mental states as to genes that make brains that have them (cf. N.K. Humphrey, 1979, p. 60: 'if a rat's knowledge of the behaviour of other rats were to be limited to everything which behaviourists have discovered about rats to date, the rat would show so little understanding of its fellows that it would bungle disastrously every social interaction it engaged in'). For survival purposes it is as important to discern functional organisation of the whole as it is to descry the movement of parts within. Because ours is a world of purpose quite by chance, our only chance is to see purpose. Organismic and teleological thinking is as much a crutch as it is a shackle. As long as it is human brains that do science, there will be a special place within science for the human; and this in spite of the long drawn out cosmological fact that the physical brought forth the human and social.

In summary, our view of a scientific theory recognises that it

(i) may have a limited domain of application;
(ii) may not be reducible to any other theory (such as physics), and may thus deploy concepts that are *sui generis* and that serve a particular kind of description and explanation;
(iii) may be non-deterministic;
(iv) must be causal in the weaker sense canvassed above;
(v) must itself have, and be produced and modified by activities that have, a certain logical structure.

3 REALISM

The philosopher's problem of the external world' has been central in the history of epistemology. What relationship do our sensory impressions bear to the 'external' world of perceived objects? The idealist dwells in the closed and cut off world of the impressions themselves, regarding objects as at best logical constructions from the stream of (inner) experience. The realist lives actively among the objects as one of them, albeit with special causal powers. The principle of objectivity drives the realist outside himself, to arrogate a God-like view of his own humble position as interactant.

Realism embodies a constant attempt to transcend the limitations of one's own subjective point of view by seeking critical agreement with others struggling with the same predicament. (This, of course, brings its dangers. In constantly trying to stand in someone else's shoes, one can easily find oneself in a forced march.)

It has two strains that produce hybrid vigour, even though the pure strains may be viable on their own. The first strain is that of *ontic* realism, implicit above in the view that observers, especially oneself (for indeed the skeptic's predicament, which he tries to convince us of, is very much a reflexive one), are just objects among others, and especially among those one perceives, including other observers. This strain of realism is thus an attempt to overcome what might be termed subjectivity of substance. The second strain has as its ruling metaphor the voyage of discovery. That is, we discover truths that obtain independently of our own intellectual activities and sensory probings. Any statement that is well-formed within the language of our theory is objectively determined as true or false, with that truth value awaiting our possible future discovery; but even if we never happen upon it, and even if we possess at present no method by which in principle one could discover that truth value, the 'facts out there' determine the statement as either true or false. The resulting theory of meaning, according to which every sentence is determinately true or false independently of our capacity to discover which, ordains that classical logic should govern inferential reasoning. It is the strongest possible logic that one can deploy, short of precipitating contradictions from consistent premises. It is *prima facie* the best logic to use as an instrument of criticism in the Popperian fashion. For Popper (1969) enjoins us to formulate bold conjectures consistent with, and possibly even implying our extant data; and then to subject them to the strongest possible tests, by drawing conclusions from them in the form of experimental predictions that can then be seen to be true or false once the experiments have been performed. The best way to do this is to employ the strongest possible logic; and that is classical logic.

The two strains of realism, treating of objects and objectivity, are combined in what has recently earned the label '(strong, critical) hypothetical realism' (Vollmer, 1975). Its ontological stance and epistemic attitude have been adopted by evolutionary epistemologists as most congenial to their project. That project is to account scientifically for the existence of cognisers, and in so doing to establish cognition's credentials in the face of old philosophical doubts. It is manifestly no 'first philosophy' designed to allay skepticism before the empirical enterprise begins. Rather, it builds on existing scientific theories, 'naturalising epistemology', making philosophy continuous with our scientific continuation of common sense.

The so-called 'argument from illusion' is frequently deployed against realists in the ontic sense. Such arguments point out the frequent mismatch between the 'real' objects and our impressions of them: the brown ellipse I see is the circular penny out there; the stick plunged into water appears bent. These phenomena teach us no more than the subjectivity of our senses; and realism of the second strain naturally seeks a way out of the predicament by emphasising how the phenomena vary across a multiplicity of viewpoints. Concerned as we are with relative constancy in our posits, we stand by our pennies, and explain tricks of light. Ironically, this rush to get at the real produces extremes such as Eddington's denial that the table before him was at all what common sense told him it was like. According to modern science, concerned exclusively with the real, it was made up mostly of empty space and whirling sub-atomic particles; the table had no colour, and was hardly to be called solid. This, however, is reductionism run amok. Not only does the reductionist in this case maintain the possibility of reduction to the level of physics, but he also denies the subsequent possibility of ever using the higher level vocabulary thus reduced in anything like its original sense.

The realist simply does not need to burn his bridges behind him. All he need do is construct further theories to account for how his constant posits map to the evanescent stream of experience. The first and obvious step is a causal theory of perception, and a treatment of cognition in general as grounded in physical structures inhabiting the very world to be cognised. Posits must fit appearances because of causal interactions between posits, some of which are appeared to posit-ly. It is only to be expected that the crudest points of fit in the present state of physical and cybernetic theory will be the philosophical chestnuts from the mind-body problem. Raw feels and qualia, it is thought by many, simply cannot retain their vivid phenomenological status or character as science goes more and more public, regardless of whether it is reductionistic. *My* pains are *mine*, and no other observer can get into my shoes on that one. Nor can I get into his. Realism's response, which phenomenalists no doubt think too pallid, is simply to leave the 'innerness' of those experiences, perhaps even the experiences themselves, entirely out of the field of fit. The appearances our posits excite within us must be describable in a common language before realist theory can map activity of posits to stream of experience. The subjective and misleading experience of the submerged stick as being bent is still describable in a common language dealing, ironically, with physical posits (albeit of common sense). Realist theory accounts for the publicity of such misleading subjective experience by appeal to the laws of refraction and the operation of the human eye. But it is of course senseless to attempt to frame public criteria according to which we would know when we had reached subjective agreement about pains, as opposed to 'appearances from this position as of a bent stick in the water'. The phenomenal clamour of inner experience drives the realist to the sanctuary of behaviourism. He no longer seeks to map pains in all their innerness to the causal activity of our physical posits, some of them housed in sensitive surfaces. Rather he seeks only regularities in total observable behaviour, in which pain states feature somewhat more pallidly as functional states of an organism. Gone is the metaphysical yearning that some observer should be able to know exactly what one is feeling. All communal life demands individual sacrifices; and the pursuit of common knowledge entails the privacy of pain.

But viewing pain states as functional states in the behavioural programme of an organism is no mean task. In the case of human beings the cybernetic interstices and transitions between them are formidably complex. To locate a state within the 'programme' for a human being as one of 'pain in the left shoulder' is to specify myriad complex ways in which the human being will behave in response to various stimuli, including linguistic ones, and how his other dispositions (such as to play sport or make traffic signals) will be affected. The realist subscribing to functionalism in psychology need not worry unduly about accusations that the phenomenalist levels against him, to the effect that on his (the realist's) understanding of matters pains are utterly divested of their significance and reality to their possessors. For this is to underestimate the gargantuan task that the functionalist sets himself, namely that just described of locating pain states within the whole behavioural and motivational programme of the organism.

The same holds for the perception of colour. The mistake of thinking it makes sense to wonder whether your green is my red and vice versa still exercises a strong hold over the philosophically non-reflective. The gap between subjective appearance and physical reality is nowhere better illustrated than in the paradox of the colour circle. Visible light has wavelengths falling within a linear interval. Various colours correlate with various wavelengths. But what is remarkable is that the colours for the two extremes of this interval seem subjectively to merge into each other, as do colours with closely adjacent wavelengths elsewhere within the interval. So subjectively the colour spectrum closes into a circle. But this does not imply that a causal theory of perception combined with a functionalist account of mind will never capture the true nature of our subjective sensation with regard to each particular colour. For we all agree on the closure of the colour circle. All the colour discriminations of which one is capable can in principle be disclosed in observable behaviour. Much of that behaviour is linguistic. We agree on colour attributions in a common language, and articulate the same calculus of similarities and differences. This observable behaviour is the only legitimate point of entry for the realist theorising about colour perception. One can do no better than perform behavioural triangulations; there is no more to the 'innerness' of my colour sensations than is conferred upon them by virtue of their place in an incredibly rich framework of interlocking behavioural dispositions. The linguistic geometry of my colour terms gives colour a station in my mental life.

We have thus far been concerned with the problem of subject and object, appearance and reality. We have distinguished ontic realism from the objectivity of truth (our two strains of realism). We have seen the pressure for publicity of viewpoint as tending to exteriorise, for good philosophical reasons, the allegedly inner and private character of the mental. Crude or naive behaviourism might well have been the Achilles heel of realism; but sophisticated behaviourism, once developed into a full-fledged functionalism, would arguably do scientific justice to the mental while anchoring it in the physical.

4 KNOWLEDGE AND BELIEF

We turn now to the question as to how we can come by knowledge in the areas in question, what methods we may employ to acquire it.

And here one must immediately alter the terms of philosophical discussion so far. Knowledge, conceived as justified true belief, has remained an untarnished ideal despite intervening scientific revolutions sloughing off the false pictures painted in the past. Knowledge in the physical world at best comes in little nuggets in the here and now. That this fly is in this bottle, that the owl is there as dusk is gathering, that that T-shirt has NN printed on it, that this is a hand before me, are the sorts of things, barring crazy philosophical objections, that I can definitely know. But that there have never been and will never be any hermaphroditic dodos? That all metals expand when heated? That the sun will rise tomorrow? About such matters one can only hold beliefs, beliefs coming in different strengths or degrees, and depending for their strength in complex ways both on what we do know and on what else we believe. Moreover, once we begin to acknowledge how theory can influence perception and its reports, and how the description of experimental results (the supposedly singular items of genuine knowledge) involves theoretical terminology from the theory being tested; and especially if we also start heeding some of the crazier philosophical misgivings about impostor natural kinds; then we begin to surrender strict knowledge claims throughout, and cleave instead to belief. Popperian methodology is then a filter for the rigorous screening of candidate beliefs, mainly in universal form. It ensures the gradual promotion and possible sudden demotion of beliefs on the scale of degrees.

Popper himself says little about the sources of new hypotheses; he confines himself to the logical structure of the activity of testing them. By contrast the inductivist tradition that he so rigorously opposes takes itself to have a lot to say about the sources of our beliefs. Various rough slogans give it its character: the future will resemble the past, like causes produce like effects, things structured similarly will behave in similar ways. Very often its flavour is imparted by a motley of examples: the sun will rise tomorrow, because it has 'always' done so in the past; all emeralds are green, because all we have seen so far are green; all ravens are black, for similar reasons; all metals expand when heated, ditto; and so on. Induction became a 'scandal for philosophy' only because it too was seen as exercised in a realm of supposedly apodeictic knowledge claims. Of course, there is no logical guarantee that if events of type A have always been followed by events of type B then future events of type A will be followed by events of type B. Hume was quite right to point out that inference from any number of singular claims will always fail to secure the truth of the corresponding universal claim. But he thereby made only the most banal of logical points. The problem about the philosopher who points out the logical possibility that tomorrow the sun might not rise, or that my first cue shot will smash into the reds and stop there, leaving them undisturbed, is that he is preaching to the converted. The Bayesian probability theorists among them have been aware of such possibilities all the time, but these possibilities have proved themselves more and more remote as past data have come in for the adjustment of life's betting odds. Hume's problem is for the Bayesian simply a matter of never being able to bet at arbitrarily low odds. Induction may be a scandal for the apodeictic addict, and gambling for monetary gain will always be (for the gambling houses aren't in business for nothing); but that we need and employ induction in the greatest gambling contest of all is no scandal whatsoever. The realist has faith in the existence of causal mechanisms, no matter how far removed from the surface these may be. The crude uniformity postulate of naive inductivism must give way to a variety of such mechanisms, competing, interacting and superposed. Their assumed existence is the missing premise in all Bayesian inductions to strengthened general beliefs from the discovery of confirming instances. My strengthening general belief that all emeralds are green is premised upon the existence of mechanisms for the refraction of light in crystalline structures with a certain chemical constitution (and perhaps also on the existence of complex neural mechanisms ensuring the association of the word 'green' with light of the appropriate wavelength). Moreover, Popperian testing is a matter of doing one's level best to engineer circumstances in which the underlying mechanism may be suspended, interfered with or thwarted without doing too much violence to the *ceteris paribus* clause implicit in *any* law that is to have applications, be it produced by a naive inductivist or sophisticated falsificationist.

From the standpoint of a realism in the sense described above, imbued with the pragmatism that replaces knowledge with a spectrum of degrees of belief, we attain an interesting perspective on the two most important responses to Hume's original problem. Popper, seeking knowledge in the strong sense, and seeking a methodology free of deductive fallacies, resorted to falsificationism, a sparse normative frame on which a great deal of suspect inductivist science could be racked. Kant, on the other hand, never once doubting the legitimacy of causal or inductive inference, made the fault lie not in the stars, but in ourselves. Causality was reckoned to the fundamental categories of thought, those 'a priori' concepts that are the conditions for the very possibility of experience. We cannot merely be making logical mistakes in descrying causal relations among the objects of sensory experience. For, according to Kant, unless we do the latter, unless we spontaneously bring causality, as a concept of our understanding, to bear on the manifold of sensory intuitions, we do not produce experience as such. We would instead be left with 'blind intuitions', mere confused ('verworren') data, that fall short of knowledge for want of conceptual organisation, or categorisation.

The problem, however, with this Kantian account is that it is not clear just how much causality, or which causal relations must be 'read into' or 'imposed upon' the manifold of sensory intuitions in order to convert it into experience issuing in knowledge. Many a far-fetched causal net may be cast by overly spontaneous understanding. All that Kant appears to establish is the need for some imposition of order, but not the need for any one rather than any other. How can we be confident that the order we impose is the right one?

The pragmatic Kantian can happily reply that we can't; we just get by as best we can. The pragmatic Kantian is one who accepts Kant's account of intuition and understanding; of space and time as *a priori* forms of the former, and of causality, substance and so on as *a priori* concepts, or categories, of the latter; but so alters the account of the operations of the understanding that the categories must come into play not only to constitute experience that supposedly leads to *knowledge*, but rather to constitute experience that is to lead to any *beliefs* at all.

Causality for the pragmatic Kantian then becomes a condition for the very possibility of belief formation and conditionalisation. The latter is the Bayesian process whereby degrees of belief are altered in the light of incoming experience. For Hume, induction, or the causal way of thinking, was a mere habit of mind; for the pragmatic Kantian, by contrast, it is absolutely constitutive of mind. For Hume, the mind to causal thinking is as my finger to its habitual tapping. For the pragmatic Kantian, it is rather as my finger is to its muscles, tendons and bones. My finger can shed its habit of tapping on the table, without thereby ceasing to be my finger. But the mind cannot shed its habit of causal thinking, without thereby ceasing to be a mind. Mind, in other words, has as part of its 'essence', the formation of beliefs and alteration of degrees of belief in the light of experience; experience which in turn consists in sensory 'intuitions' having been categorised in a causal fashion.

5 EVOLUTIONARY EXEGESIS OF KANT

At this point in the synthesis of philosophical views evolutionary considerations can make a telling entry. Kant had said the following about the concepts of time and space, as forms of intuition, and the categories, as concepts of the understanding:

> To seek an empirical deduction of either of these types of concept would be labour entirely lost. For their distinguishing feature

> consists just in this, that they relate to their objects without having borrowed from experience anything that can serve in the representation of these objects . . . in view of their subsequent employment, which has to be entirely independent of experience, they must be in a position to show a certificate of birth quite other than that of descent from experiences.

Now by deduction', it should be remembered, Kant meant the explanation of the way in which concepts can relate to their objects. Kant says that it does not matter that one needs sensory input in infancy to awake and stimulate the development of the *a priori* forms of intuition and concepts of the misunderstanding; for their distinguished mark was 'that they relate to their objects without having borrowed from experience anything that can serve in the representation of these objects'. They cannot be accounted for in the empirical fashion he describes Locke as favouring. Their Deduction 'is not to be looked for in any such direction'. The wrong direction in question here, of course, is the ontogenetic one. Ironically, Kant then uses a biological metaphor. We must, he says, if we wish to be justified in attributing to them a sense and definite meaning, seek for them 'a certificate of birth quite other than that of descent from experiences'. But Kant's own '*Geburtsbrief*' simply attributes to the understanding a constitutive causal caprice. He may be right, at the level of philosophical analysis, that that is how a mind works in apprehending order in the world, and arriving at knowledge (or as we would prefer to say, beliefs) about the objects of sensory experiences. But the main problem for the theory of knowledge is not to discover how the mind works (although that, of course, is important) but rather to explain how its working in the way it does produces knowledge (or reliable beliefs) about the world we inhabit. *That* is the '*Geburtsbrief*' one is after. Kant named the problem himself: how can subjective conditions of thought have objective validity?

The famous answer given by Lorenz, on which the whole of evolutionary epistemology is based, is that what is ontogenetically *a priori* is phylogenetically *a posteriori*. Life, says Lorenz, is an '*erkenntnisgewinnender Prozess*'. But in the light of our recommended shift from concern with knowledge to concern about belief, it is time to replace this phrase. Life is not an *erkenntnisgewinnender Prozess*; it is a *glaubensgewinnender Prozess*. Successful cognisers enjoyed an evolutionary advantage. Those that saw, smelled, felt and heard aright were favoured; while the radically off-beam realists, along with the skeptics if ever there were any, were wiped out. This applies throughout the phylogenetic order. Every organism possesses a '*Weltbildapparat*', an apparatus serving up representations of the environment, at least in those respects that are important for reproduction and survival. The most fundamental and reliable features of the world were slowly 'wired in' by natural selection, thereby producing more efficient processors of information and hence cognition. Those most fundamental features are, of course, the Kantian *a prioris*: space and time as forms of intuition, and the various categories of the understanding. The latter list might be updated since Kant's day; he was very much dependent on Aristotle, for the quaint reason that he thought that the categories had somehow to correspond to Aristotelian forms of judgement. A more modern list we would suggest is the following: the concept of substance, the notion of a continuant; concepts of identity by various sortal criteria; the notion of event and of cause.

Another important category on Kant's original list, and one which surely must remain even though contemporary evolutionary epistemologists have hardly mentioned it, is that of 'Community (reciprocity between agent and patient)'. This is the category whereby a combination of things into a whole is made in thought: '. . . for one thing is not subordinated, as effect, to another, as cause of its existence, but, simultaneously and reciprocally, is co-ordinated with it, as cause of the determination of the other (as, for instance, in a body the parts of which reciprocally attract and repel each other)'.

This category of *functioning aggregate* as opposed to haphazard collection of disparate parts, or as opposed to inert class of elements without unitary significance, is surely of the utmost importance in the recognitional capacities of an intelligent being. To recognise a functioning whole as such, be it an organism or a special unit, is, of course, not necessarily to espouse an holistic or anti-reductionist philosophy, nor to call for the latter in the explanation of the mechanisms underlying such recognitional powers. The guiding tenet of methodological individualism is that functioning aggregates precisely owe their identity and causal powers as such to the causal clustering of their components, and to nothing more. Obviously the final explanations to be had from a completed science devoted to this methodological tenet will be fearsomely complex. But so too would be the proof of any reasonably deep mathematical theorem if it had to be spelled out in the fullest possible detail, without using any abbreviatory definition. Reduction always shatters the semblance of outer simplicity, of unitary coherence, of social cohesion, splintering the integrated system to be analysed. It pursues all those interactions that ultimately constitute the whole phenomenon, interactions which are not necessarily constitutive of the phenomenon of the whole as categorised by the cognising intellect upon first acquaintance. Synthesis is a recognitional *fait accompli*, the instantaneous product of a presently mysterious intellect. We can reckon the category of *Gemeinschaft* as basic to the recognitional powers of human beings (and others (such as rats?) – cf. Humphrey, 1979) without in any way committing ourselves to a super-organicist ontology for (social) science. What partly makes science so difficult is precisely that its formal analytical methods can run counter to the spontaneous modes of recognition governed by the Kantian categories. The latter may one day be reconstituted in a synthesis yet to come, when neuroscience and cybernetics finally yield a workable model of human cognitive capacities. Until that time, we shall be in a no man's land, like the child who has dismantled his best loved toy and does not yet know how to put it back together again.

We mentioned in chapter II the possibility that our brains might have been shaped by evolution spontaneously to recognise the social significance of actions as much as to see events in purely physical terms. This suggestion prefigured the present recommendation to reckon *Gemeinschaft* to a modern list of Kantian categories; while yet, of course, not surrendering to the analytic or reductionist methods of modern science when attempting finally to understand the holistic and wholistic competences of cognisers.

Modern science has taken, and no doubt will continue to take us, along a hard won route to an understanding of a much wider range of phenomena than ever had selective significance for our hominid ancestors. Long before Euclid invented the axiomatic method in geometry, before ancient societies developed the number system, before the Presocratics speculated about ultimate categories and constituents of the universe, before Galileo invented the experimental method in physics, before Bacon fashioned the canons of inductive science, human beings were surviving pretty well. Their intellects were spontaneously applying our neo-Kantian categories to a flood of everyday sensory experience. Their resulting belief systems might, of course, be regarded today as ever so skew – as full of superstition and quaint theory – but they were after all simply earlier passengers on the Neurath raft which each generation of humanity inherits from the last. The community-wide sharing of individual beliefs, the possibilities opened up for the transmission of the new beliefs of a particularly gifted intellect, enabled the long hard Bayesian trek to our present epistemic vantage point.

A great deal of that hard-won theory has helped men drastically to alter the selective potency of certain environmental conditions. Disease has been reduced, infant mortality lowered, access to warmth, food and shelter improved and so on. But the Kantian account of sensation and judgment via the forms of intuition and the concepts of the understanding was no doubt as valid for our ancestors of a few thousand years ago as it is for us today. Or was it? One of the central claims of evolutionary epistemology is that our cognitive capacities, being grounded in neurological structures that are themselves based on the action of genes, have been phylogenetically shaped in such a way that Kantian *a prioris* are ancestral *a posterioris*. We have emphasised above how fast natural selection can operate to produce changes in social behaviour. Is not the possibility likewise open that even over the last five thousand years, during the period from the inception of reflective philosophy and experimental science to the present, genetic changes may likewise have considerably altered man's Kantian apparatus? We think not. Although there may have been some change, it is difficult to make out a case why particularly good scientists should have been reproductively more successful. The important shifts in cognising since the birth of culture have taken place at a more formal and symbolic level. But even Einstein could not imagine what it would be like to inhabit some of the spaces he contemplated in his theoretical work, what the field of sensory intuition would be like, or what intuitive grasp one would have of causal nets and interactive ensembles. Kant may have been discredited by the discovery of non-Euclidean geometries, and the discovery that some of these provide 'better' descriptions of the spaces contemplated in relativity theory. But let us not consequently dismiss spatiality as such as a *sine qua non* of sensory intuition. Every geometry of any likely value for a simple physics at least delivers a notion of betweenness, and a metric according to which some points are nearer to one than others. That much at least of space as an *a priori* form of intuition must be common to the ancient Greeks and cosmologists for generations to come.

That mankind could have 'discovered' that the fundamental properties of space-time are *not* as we immediately intuit them need not disturb the evolutionary epistemologists. The relevant domain of selection has, to all intents and purposes, been Euclidean. While also there has not been selection *for* the ability to do relativity physics, that ability is nevertheless the result *of* selection for other general abilities. And even if the contrast between our present, scientific account of space-time and that delivered by unschooled intuition cannot be attributed to genetic change, then at least one can still appeal to the forces of cultural evolution to account for our intellectual desertion of geometrical *a prioris* while yet our sensibilities labour under their archaic tutelage.

Evolutionary epistemologists have concentrated most of their attention on the notion of cause, with Riedl (1980) and Wuketits (1981) prominent among them in enquiring after '*die Ursachen des Ursache-Denkens*'. They seem, however, to have neglected the notions of space and time, as forms of intuition as opposed to concepts of the understanding; and not to have dealt very well with the other categories, such as *Gemeinschaft*. Some sloppy expositors among them even call space and time categories, with no apparent exegetical qualms; others maintain straightaway that Kant was interested in *structures*, take these to be neurological structures, and immediately launch into their scientific stories of how we come to possess them today. The value of their own contributions is not to be underrated by these remarks of caution. One small contribution we would like to make to the growing discipline is that of slightly re-focusing Kantian spectacles so as to connect evolutionary theorising more effectively with the core doctrines of the *Critique*. In some places these doctrines will emerge distorted beyond recognition.

First, some reasons why one might dismantle the strong distinction between space and time as forms of intuition and the categories as concepts of the understanding. We think some case can be made for running these together, by considering more closely how one perceives.

First, Kant's notion of space and time are too monistic or unitary. Kantian space incorporates all the mathematical properties of Euclidean space – infinity, isotropy, homogeneity, continuity, connectedness, along with the metrical properties that allow one to speak of angles and to think of a unique line parallel to any given line and passing through a given point off the latter. This is a vulnerable unity to hold up as an *a priori* form of intuition. The multiplicity of modern geometries enjoins us to make a more discriminate selection of basic spatial *concepts* such as betweenness and relative nearness that are shared by all geometries. No longer then is space an *a priori* form of intuition; rather, certain spatial concepts become *a priori* concepts of the understanding – that is, are reckoned to the categories.

Let us consider how we must change the Kantian account of sensory awareness and intellectual organisation. Kant considered an essentially two-step process producing knowledge. First, the manifold of sensory intuition impinges on our awareness as spatially organised and as occurring at or enduring through a short moment of time. (Kant also speaks at one point of this manifold being synthesised by the imagination; whether to count this as a separate step, or as part of the first, is not clear.) The synthesis of the manifold, however, does not yet produce knowledge. That comes only when intuitions are united by concepts in the understanding.

This tandem process should now be welded into one. It is difficult to understand how a manifold of sensory intuition can be spatially organised without appealing to, or already applying, certain sortal concepts. If sensory datum A is to be deemed at the first stage as lying further from the perceiver than sensory datum B, something more than two-dimensional spread and texture within the visual field must be involved. *A* will have to be taken, say, as a datum *of a distant house* rather than as a datum *of a nearby architect's model of a house*. The spatial organisation of sensory data within the manifold already requires an act of understanding, or of visual interpretation; concepts carving out objects must already be applied in bringing that spatial order to the manifold, order which is a precondition for the very existence of the latter. Moreover, the spatial ordering proceeds by applying a small stock of spatial *concepts*, now on a par with the sortal concepts that must combine with them to effect that ordering. Thus interpretation begins at our sensory surfaces, at the very moment incoming stimuli begin to be filtered along neural pathways (to see matters in a neurological way for an intrusive moment).

In so structuring the manifold through spontaneous cognitive activity, the mind is not all the time making judgments, not even in an inner sense. It is rather synthesising mental representations on which dispositions to judge aloud, or to act as if a certain judgment is held, are based. The interpreted sensory manifolds of different times and vantage points must further be synthesised by cognitive processes that preserve a sense of one's own position relative to perceived objects, and one's own sensory orientation in space. The accumulating 'data base' formed as time goes on consists of data that are already highly interpreted, or processed by means of concepts possessed by the cogniser. It forms, as it were, his 'atomic diagram' of his occupied stretch of reality.

It need not disclose generalisations or universal claims about all objects of such-and-such a kind, some of which lie outside that stretch. But that such generalisations are made, and very often spontaneously and unconsciously so, is a fact of cognitive life. From the singular roots of the perceived here and now, and subsequent there and then, a great tree of speculation grows. In a creature such as man, it has a fundamentally linguistic organisation. Newer categories of thought, of logic and mathematics perhaps, have to be applied to trim and control this burgeoning growth. It is no historical accident that mathematical and logical reasoning became more conscious as theoretical endeavour increased its reach and spread.

The foregoing revision of Kantian doctrine breaks down the divide between forms of intuition and concepts of the understanding. Perception is conceived as a more unitary and interpretative process. The result is not always knowledge, but rather dispositional belief. The basic information in the manifold supports a set of linguistically coded beliefs, which is constrained both by logic and by rational processes of belief modification (the sophisticated modern cousin of the category of causality discussed above). We believe nevertheless that it is in keeping with the spirit of Kant's enterprise:

> We shall therefore follow up the pure concepts to their first seeds and dispositions in the human understanding, in which they lie prepared, till at last, on the occasion of experience, they are developed, and by the same understanding are exhibited in their purity, freed from the empirical conditions attaching to them.

There is no activity without a substrate. Minds must be embodied. This is a fundamental tenet of materialism, one accepted both by us and by the evolutionary epistemologists to whose work we are addressing ourselves. We would argue that we should tacitly disregard the Kantian distinction between forms of intuition and concepts of the understanding, opting for a more conceptual permeation of the perceptual process. Moreover, we mentioned above that some evolutionary epistemologists slip all too readily into talking of neurological structures when putting Kantian doctrine into an evolutionary perspective. How can they be justified in doing so?

The answer, we think, is as follows. Short of a thoroughgoing dualism, or a functionalism insisting on absolutely wild, erratic, random and chaotic physical realisations of functional states in different brains, there is a great deal of attraction in the idea that there is a considerable homology between mental and neural states (*pace* even the mild functionalist). The revised Kantian account above of the phenomenology of perception and of judgment is one that should be congenial to anyone working in the tradition of cognitive psychology. The categorical basis of concept and belief formation, with the latter process itself, and the extent to which the categories and the constraints on those processes are innate or given, or innate but maturing, is a legitimate subject matter for the investigator interested first and foremost in the functional organisation of an intentional system like the human mind. But even the cognitivist acknowledges – or should acknowledge – that intelligence and sentience do not reside in a programme, but only exist when the programme is run. Mind must find a body; it is only the embodied mind, or animated body, that is the proper seat of intelligence and sentience. Given this, it is not implausible that the given must be 'wired in': the innate must be laid down in cerebral circuitry, be it biological or electronic. Moreover, if the resulting cognitive capacities have given rise to successful behaviour favoured by natural selection, then these capacities became more widespread because of the proliferation of the genes that coded for the physical growth of the neurological structures embodying the innate givens. Materialist convictions thus underlie the transition from *a priori* Kantian categories (and we would add, constraints) to underlying neural structures. Only if this is so can they be thought of as having evolved.

This now brings us to a further point of exegetical strain on the part of evolutionary epistemology as far as Kant is concerned. Kant was certainly agreeably relativistic about other species:

> What objects may be in themselves, and apart from all this receptivity of our sensibility, remains completely unknown to us. We know nothing but our mode of perceiving them – a mode which is peculiar to us, and not necessarily shared in by every being, though, certainly, by every human being. With this alone have we any concern.

But evolutionary epistemologists are concerned with a lot more than just our own human cognitive capacities. In so far as they wish to account also for human ones, they do, of course, try to bring their theory to bear on Kant's doctrines. But when mapping human understanding as an evolutionary product, and trying in their interesting fashion to ground individual *a prioris* as phylogenetic *a posterioris*, they face a further problem in Kantian doctrine. For Kant's analytic was to be appreciated as follows:

> Transcendental analytic consists in the dissection of all our a priori knowledge into the elements that pure understanding by itself yields. In so doing, the following are the points of chief concern: (1) that the concepts be pure and not empirical; (2) that they belong, not to intuition and sensibility, but to thought and understanding; (3) that they be fundamental and be carefully distinguished from those which are derivative or composite; (4) that our table of concepts be complete, covering the whole field of the pure understanding.

Now once the Kantian categories have been naturalised by evolutionary epistemology, it will be all too easy to lose sight of Kant's transcendental delineation of the *a priori*. If, in this naturalised account, the *a priori* is to be regarded as what is innate in our cognitive inheritance through evolution, then it once more becomes an empirical question just *what* categories of thought and constraints on the process of forming and modifying concepts and beliefs *are* innate.

What the evolutionary epistemologist gains on the scientific swings, he loses on the analytical roundabout. We may sadly learn that more *a prioris* have been embodied in our brains than is desirable in an ideally rational being. Or, *if* we maintain human rationality as the only kind of rationality there is, and match the richer innate repertoire of categories and constraints (or perhaps licences) that evolutionary epistemology may uncover, with transcendental deductions that go through only because we are prisoners of our own perspective; *then* that will be a Pyrrhic victory indeed, a *Selbstbestätigung* of the purity of a possibly far from ideal form of rationality.

Bad intellectual habits can be learned; or they can be inherited wired in. There is considerable evidence that some of our habits are – according to highly reflective criteria whose formulation attests to our extraordinary ability to be reasonable in spite of ourselves – simply bad; moreover, one cannot rule out the possibility that they are innate. People leap to statistically ill-advised conclusions about future throws of coins and dice on the basis of very short initial sequences of throws (Riedl, 1980, pp. 56-7). They subsequently fail to revise their beliefs in the light of fresh data in the way ideally required by the Bayesian rule of conditionalisation. People also make some of the simplest and worst fallacies in the logician's book (such as affirming the consequent) in their daily reasoning (Evans, 1982). How is it that one can on the one hand excogitate normative canons of thought, and on the other apparently be capable of violating them so readily, so spontaneously, in everyday thinking? Do we have a genuine Kantian competence innately wired in, which careful and reflective mathematicians and scientists are more attuned to, with other peoples' aberrations explicable as the result of 'interferences' in some performance model? Or are we wired in an evolutionarily makeshift way, producing only a satisfactory statistical outcome for the survival of our genes? He who espouses an evolutionary account of our cognitive structures as the best explanation of the *a priori* is an optimistic Kantian indeed if he does not carry over to the mental the unsentimental lessons evolutionary theory teaches us about adaptation at the morphological level. We know now that adaptation is imperfect; that vestigial organs remain even when they can be encumbrances; that human embryos have gills, and that even adult organisms are lumbered with design features of the past. So why not the same in the case of mental organs and their operations? Evolutionary epistemology can be expected to render revealed reason as all the more vulnerable, all the less ideal, for being the product of our evolutionary past.

6 BIOGENETIC STRUCTURALISM

We mentioned above that not all the credit in recent evolutionary thinking about our cognitive functions is due to the Austro-German school. The American anthropologists Laughlin and D'Aquili (1974), working apparently quite independently of this school, have produced remarkably similar ideas. They are clearly on to something like Lorenz's idea of open genetic programmes being used to compile advanced behavioural and learning patterns. These open programmes require an input of information before 'hiving off' routine recognitional and motor skills to more settled sectors of cerebral industry. Imagine the grey matter of a new-born infant. If one accepts that the physiological changes that endow it with mature cognitive capacities amount to the formation of neuronal connections to be insulated by a myelinalised layer, along with the shedding of many brain cells, one is easily impressed by the metaphor of cerebral crystallisation, the sharp definition of topological possibilities already somehow fuzzily there as labile a prioris. Laughlin and D'Aquili mention Lorenz not at all, and Kant only once, fleetingly on p. 3. Yet on p. 10 we read that

> A neurognosis (neurognostic model) is a specific initial form in which reality is experienced. It is a model, as it were, that is minimally isomorphic with reality. *It may be cognitive or affective*, or both. [Our emphasis – cf the account of the appetitive and the cognitive in Chapter II above.] It is often capable of being linked to (or 'described by') some verbal utterance that itself may be simple or very complex. The verbal utterance, in turn, may have the attributes of a model of the neurognostic model. An example of one such simple formulation is 'All events in the real world are ordered in causal sequences.' This utterance is obviously not the neurognostic model itself, but it does form something of a shorthand description of the model, or a class of models.

One is reminded immediately of Kant's causal Grundsatz, and of his account of how the synthetic *a priori* truths expressed by statements of arithmetic and geometry have their source in space and time as *a priori* forms of intuition. In a later paper Laughlin and D'Aquili (1975) list the tenets of their biogenetic structuralism, which conform remarkably well to those of evolutionary epistemology. Neural models are phenotypic expressions formed by fields of neural connections located primarily in the brain. They are 'vulnerable to natural selection and reflect detailed patternings specific to the species . . . the neural structures are mutable and form an open system through the empirical modification cycle . . . comprised of somatosensory input and behavioural output'. In at least some cases the cycles of modification and elaboration are regulated by 'genetically determined ontogenesis', our criterion for heritability in the first sense, or innateness. The cycle produces stages, lineal progressions, and 'critical periods', all of which meshes with the findings of developmental psychologists such as Piaget (1952).

Laughlin and D'Aquili have attempted to identify certain cognitive or neurognostic structures with localised areas of the brain. Now the possibilities of wide cerebral diffusion of any one cognitive function are well known; but these two authors are reasonably confident about the cerebral localisation of the following 'operators':

1 the holistic	4 the binary
2 the causal	5 the formal-quantitative
3 the abstractive	6 the normative

> Neurognosis is, as the term implies, biogenetically given rudimentary information about the world, embedded in various association areas of the brain, either by the time of birth, or at some later maturational stage. The neurognostic model . . . may be viewed as something like a first approximation (or model 'seed') of phenomena in the world, models later to be modified, amplified, and even circumvented through learning . . . we contend that the neurognostic models form the initial 'crystals' around which form many other behaviour patterns that, like music, primary social group identification, and conception of repetitive time, appear to be universal, or virtually universal, among human societies. (p. 83)

Most of the neurophysiological processes obviously take place outside the bounds of awareness, including, they maintain, some of those comprising 'higher cognition' in man. This is once again Kant's spontaneous activity of the understanding in a more modern materialist guise.

There is undoubtedly a kernel of truth, be it empirical or philosophical, in these converging lines of thought on the part of ethologists, psychologists, anthropologists and neurologists. One hesitates only to see neurognosis everywhere, a brain spiked with an overly rich mixture of cognitive spirits. Kant's problem of elementariness should constantly be borne in mind also. Identifying certain 'fundamental functions' as associated with a localised area of the brain is all very well. The question to be answered, however, is 'fundamental by what criteria?'; and it should also be borne in mind that even if the allegedly fundamental operations turn out to be highly complex in cybernetic terms, as involving a great deal of sub-computation and compiling, that would still be quite compatible with our finding that a little lesion at some particular spot throws a massive spanner in the works of that 'fundamental operation', without yet shutting down the whole works. The ideal compromise would be a list of categories nicely allied with neural structures that are similar in all human brains; with the categories bearing the hallmarks of convincing conceptual analysis, able to serve as intellectual building blocks for the whole system of human thought; and with the corresponding neural structures emerging during ontogenesis from innate germinal outlines within the new-born brain. But that is a high hope indeed; and materialistically minded philosophers of mind might have to prepare themselves mentally for other possibilities. The main comfort to be drawn for a materialist philosophy of mind from evolutionary considerations is that the embodiment of mental functions within the brain might be relatively uniform across the species.

7 THE BRAIN AS AN EVOLUTIONARY PRODUCT: THE EVOLUTION OF CONSCIOUSNESS

The human brain consists of three basic portions that are functionally distinct and arose in successive stages of evolution from amphibians to man. Maclean (cf. Holden, 1979) describes it as an onion with concentric layers. The layers exercise different control functions and give rise to different capacities. These correspond roughly to the appropriate capacities of the species in those respective phylogenetic stages. The oldest and innermost component is the brain stem whose main function is to regulate metabolism. Then comes the cortex which governs important kinds of instinctive behaviour. This is the seat of our feelings and unconscious motivations or drives. Lastly comes the neocortex. This is most pronounced in man, and is bound up with thought and action, and our logico-linguistic abilities. Nevertheless, consciousness can be 'shut down' by interference with the brain stem, the oldest part of the brain. In addition to the onion structure is the walnut structure of this remarkable organ. The brain is divided into two 'hemispheres', each bound up with the sensory-motor abilities of the opposite side of the body. They are joined by the corpus callosum, a bundle of nerve connections that allows a vast exchange of information between the two hemispheres.

Neurological research is proceeding on two fronts. One is to investigate the fine structure of our grey matter, to discover how the brain cells transmit electro-chemical impulses in the network of dendritic fibres joining them up into an astronomically complex jungle gym. The other is to investigate the effect on people's feelings, emotions, speech and thought of accidental brain damage, lesions and lobotomies – that is, relatively gross disturbances of overall structure. Some remarkable findings are emerging, with often controversial philosophical implications. The well-known 'split brain' experiments allegedly revealed that patients suffering a cut of the corpus callosum manifested distinct 'personalities' endowed with different capacities that are normally integrated in the healthy and undamaged brain (Gazzaniga, 1970). The left hemisphere is more closely bound up with the logical and linguistic functions, the right with the emotions. This, of course, is a very crude summary; the experimental data have given rise to controversial interpretations. Perhaps the most sensational 'discovery', if it counts as one, is that the two hemispheres even have rival senses of personal identity, as though a healthy person is the democratic front of a gang of two (Sperry *et al.*, 1979). Apart from cuts of the corpus callosum, lesions and accidental damage elsewhere have enabled investigators slowly to piece together a crude map of cognitive functions and sensory-motor control centres. Particularly in the case of speech disorders, one finds an intricate spectrum of effects – slurred speech, telegraphic speech, fluent but meaningless speech, loss of particular kinds of comprehension, unconscious swearing that the patient cannot recall, to name but a few – that show just what an intricate ensemble of components must be involved in normal competence.

Finally, there is the phenomenon of so-called equipotentiality, or plasticity of the brain tissue. People have been discovered late in an apparently normal life with whole hemispheres shrivelled or missing – apparently able to run cognitively on one cylinder (cf. Lewin, 1980). (Indeed even 90 per cent of whole cortex has been missing in some of these cases.) The earlier in life massive damage occurs to one hemisphere, the better are the chances that the other hemisphere will provide a home for the orphaned functions. It appears that the 'lateralisation' of function revealed in the split brain experiments takes place as the brain matures, becoming more pronounced and irreversible as time goes on.

One theory of consciousness, or awareness, is that it emerges from the integration of control functions, as a monitor of subordinate monitors, arising from the growing competition among different drives and inclinations. Consciousness 'is there' to cope with the new and unexpected, and to take appropriate measures; thus when a sudden pain, say, abruptly impinges upon it, or a sudden danger arises in the visual field, we are most highly conscious of it – even though, ironically, we may 'forget ourselves' in the ensuing agony or panic. Events can thus rise up and demand our conscious attention; or we can consciously direct our attention to less clamorous phenomena – to planning the day's work, examining one's motives, appreciating works of art and so on. Some of what are sometimes (perhaps unreflectively) called heightened states of consciousness are also, paradoxically, some of our most 'animal' states – the glow of satiation after a good meal, or after making love. What we are normally least conscious of is the routine, the humdrum, the fully predictable, the movements at which we are highly skilled. One ceases to be conscious of the disposition of furniture in one's favourite room, noticing a chair only if it has been moved; one can drive a car, changing gear, braking, turning, accelerating, while one's mind is 'on other things'; one shaves in the morning while fully absorbed with the problems of the day ahead. 'Consciousness raising' is a fashionable phrase among political groupings whose members feel particularly oppressed by the humdrum of repetitive and undemanding work. Consciousness is provoked or raised by the new and unfamiliar, by the exciting or threatening. And according to some psychologists, when something is too frightening or threatening, it is sometimes repressed, with neurotic consequences.

It will be a long time before considerations such as these – which any moderately reflective mind could extend indefinitely – can be brought within the compass of a single functioning 'artificial' cybernetic system, and within the scope of evolutionary explanation. But evolutionary epistemologists have made a start.

So long as one accepts that consciousness arises as a result of physiological – especially neurological – processes, or at least as a result of the concomitant complexity of organisation in functional states thus embodied (at what might be called the 'software' level of the organism), then there should be no problem in explaining why it has evolved, or been selected for. As an emergent property of physically complex living systems, it is due ultimately to the action of genes. It is therefore subject to selective pressures resulting in evolutionary changes, just as in the case of any trait that increases the organism's inclusive fitness. And the way genes for consciousness might secure their own spread could in part be to construct bodies that treated their kin as ends rather than means.

One would expect consciousness to make its phylogenetic appearance gradually, and not necessarily to appear in full-blown form with one genetic saltation. While not convinced that it is legitimate to speak of consciousness as unitary, rather than as manifest only in a variety of modifications of different sorts of functional state, for present purposes we shall allow ourselves such licence. The reader may, if he wishes, limit himself to thinking of conscious *belief* throughout, in order to extract the gist of what we say about consciousness 'itself'.

What sort of spectrum for this property is plausible? Perhaps first there is sentience, greater or lesser according to the variety of sensory channels or modalities, and their integration within the motor system. The sensory-motor system, whether in a species without a nervous system (such as Paramecium) or with one (such as a rat) is essentially a means of mediating between sensory input and motor output. The more complexity there is on both sides (and it is almost a conceptual truth that complexity can increase only in tandem) the more pressing the need for hierarchical or modular control, and the more scope there is for consciousness to emerge as a central control function, or as an aspect of states thus involved.

After the 'merely sentient' there are creatures whose behaviour invites attribution of beliefs and desires in explaining intentional actions. At this point one might be able to speak of animal awareness, something still short of consciousness. Some interest has recently attached to experiments attempting to show that chimpanzees, orang-utans and gorillas can 'recognise' themselves in mirrors whereas monkeys cannot (Gallup, 1979). Whether this is sufficient to credit the former with *self*-awareness is problematic. As far as we can see, the conclusions to be drawn from these recognition experiments are more modest. They reveal at best that the apes have a rudimentary understanding of the fact that reflections are just that – *reflections*; and that in the circumstances concerned these were reflections of the bodies before the mirrors, namely (and here we speak as observers) their own bodies. Thus they display a cognitive skill much like that of an ape who could, say, accurately spear a minnow nibbling his submerged toe, by making the required allowance for the angle of refraction.

Conscious beliefs we would be prepared to acknowledge once the behavioural repertoire allowed for the attribution of higher order beliefs and intentions – beliefs about others' beliefs, intentions to change them, and so on, regardless of whether structured language is present among the creatures concerned. Mellor (1980) has indeed identified conscious belief as belief about one's own beliefs, but the sophisticated logico-semantical problems will not detain us here. It is roughly at this stage that the ability to imagine oneself in future or counterfactual situations, to plan, project and plot, attains decisive evolutionary significance. Anticipation, foresight and deliberation are the monopoly of complex behavers, whether or not they yet be language-users. As we shall see in the next chapter, at least one conceptual route to meaning provides for a complex behavioural medium in which language is nourished. Now whether or not semantically structured language develops or could develop before speaker and hearer are able to form complex mutual beliefs and intentions, we take it that one would be on firm ground in attributing self-consciousness to agents among whom such a belief-intention mechanism was at work. For the mechanism, whether or not enabled by or enabling language, depends (after Grice, 1957) on the communicator's having an intention to induce in others a belief about *his own* beliefs. Since beliefs are states that are attributed solely in order to explain intentional actions, this entails that the communicator has a conception of himself as an intentional agent; and that we take to be tantamount to self-consciousness. Certainly, if the Gricean mechanism is there, along with a structured language with a first person pronoun system *and* attitude verbs that can be iterated, one can hardly entertain the possibility that the agents lack self-consciousness.

These theoretical or philosophical classifications appear to impose Rubicons or thresholds, 'banding' the emergent property in question. But this is only an analytic artifice. Just as sensitive and comprehensive consideration of betting behaviour would allow one to place a man's beliefs at various points on a continuously varying scale from 0 (definite disbelief) to 1 (conviction), so one might advance the view that some complex method could be developed to locate personal 'degrees of consciousness' on a continuously varying scale. Assuming genetic variability for this trait, one would then expect not only possible ontogenetic increases in personal consciousness through meditation, introspection and scientific self-understanding, but also yet further evolutionary development of the trait. Exaggerated or unbalanced forms of it, such as in hypochondria or Hamlet, might, of course, turn out to be maladaptive.

It is to be expected that a human infant, in Haeckelian fashion, would grow through the stages just described. Indeed, this is by and large in intuitive agreement with what we observe. At first the new-born babe is sentient, certainly 'registering' or perceiving features of its environment, acquiring very transient beliefs and making its first fumbling motor responses on the strength of recurring but limited desires. Then when sensory-motor co-ordination is more fully developed, and the pre-linguistic toddler plays with toys, searches under furniture, reaches up to tables, bangs spoons in bowls, one is willing to credit it with perceptual beliefs of some duration, reasonably specific and varied desires and even some long-range intentions. As every parent knows, there follows a difficult phase during which it seems that the child's tantrums are largely the result of his having quite specific desires and intentions, but being thwarted in fulfilling or executing them for want of that crucial enabler, language.

And then language arrives in a rush. At first the child will express desires by referring to himself by proper name or by means of a general term ('baby') that to all intents and purposes functions as a proper name. As he masters, say, English, his sense of self is sharpened by his growing grasp of the Gricean mechanism and by his mastery of an unspecific word ('I') that is the same for all users but of potent pragmatic specificity in context. It is tempting to view the mature sense of self as residing in this mastery, the linguistic expression of consciousness of one's own self among other selves (cf. Anscombe, 1975).

One writer has even maintained that there has been a significant change in our sense of self since the time of the ancient Greeks. Jaynes's analysis (1976) of the *Iliad* leads him to the conclusion that the ancient Greeks' mode of self-consciousness or self-representation was importantly different from our own.

> Iliadic man did not have subjectivity as we do – he had no awareness of his awareness of the world, no internal mind-space to introspect upon. In distinction to our own subjective conscious minds, we can call the mentality of the Myceneans a *bicameral mind*. Volition, planning initiative is organised with no consciousness whatever and then 'told' to the individual in his familiar language, sometimes with the visual aura of a familiar friend or authority figure or 'god', or sometimes as a voice alone. The individual obeyed these hallucinative voices because he could not 'see' what to do by himself.

This, of course, would ascribe a startling power to the combined forces of genetic and cultural evolution in breaking down the bicameral mind in but a few millennia. Jaynes, as we shall see below, also suggests that language made a relatively late evolutionary debut. It is to the problem of how language might have arisen that we now turn.

IV

THE EVOLUTION OF LANGUAGE

1 INTRODUCTION

In 1860 the Linguistic Society of Paris banned publication of the theorising about the origins of language then in vogue. A now fashionable intellectual reflex is to dismiss the subject out of hand, as too speculative and of dubious interest or value.

On the contrary, anyone concerned with the evolution of man must address himself to the problem of how he might have acquired his crucial capacity for language. Inferences about the origin of language can today draw on a vast assemblage of data and hypotheses in neighbouring disciplines. Those who belittle the value of new integrative speculation are, in a phrase of Bennett's (1976), dogmatically defeatist. In the past century we have come to understand better the structure of natural language, and have made some conceptual progress in the philosophy of language. We have learned much about our own brains and vocal tracts, and those of our primate cousins. Darwinian theory has been synthesised with modern genetics. We have attended closely to how children learn to speak. In the light of all this, we do not think it premature to put forward an evolutionary scenario for natural language.

Interest in language origin theory, (e.g. Harnad *et al.*, 1976), is now being revived after a long period of disrepute. Hardly any writer, however, has looked, linguistically, beyond unstructured lexical items. Our attention will be held far more by the enigma of syntactic structure, and how its development might mesh with the conceptual analyses of meaning and intentionality offered by philosophers of language.

We hope to show, among other things, how the phenomenon of

recursive syntax may bear on the intentional ingredient in human communication. We also say something about the structure of primate brains as well as the mechanisms of linguistic evolution. Our thesis can best be summarised as follows:

> There could be staccato talk without thought.
> There had to be thought before structured talk.
> Once established, structured talk could be mastered with less thought.
> Once mastered, structured talk makes for more thought.

2 MEANING AND CAUSALITY

It is a philosophical commonplace that we cannot learn very much about human language from the singing of birds or the dancing of bees. Systems of animal 'communication' are held to be mere signalling systems, systems for which there is some doubt as to the meaningfulness of the messages contained therein as opposed to their characteristic causal efficacy in evoking or triggering certain responses. The acts of communication that we attribute to animals are regarded by many philosophers as falling short of being *linguistic* acts.

The dance of the bees has a repertoire of wiggles and tilts and speeds. Certain features of their performance are found to be correlated with and hence to convey information about direction and quality and distance of food source. But do the movements 'convey information' to fellow workers in the hive in any more *semantic* a fashion than, say, a bruise on a child's face conveys to his mother the information that he has had a nasty bump?

At first sight, it is unclear what the answer ought to be. The bruise is a direct causal consequence of the bump. Knowledge of human physiology would enable witnesses of the bump to predict the appearance of the bruise. Similarly, detailed knowledge of bee physiology should enable one to predict that the bees' locating a food source will have as a causal consequence (even if only statistically) the subsequent pattern of wiggles and tilts in the hive.

So is it just a matter of degree? Can one argue that the bees' dance is as much a causal consequence of this earlier experience as the bruise was of the bump? Compare Armstrong: 'utterances of sentences in the communication situation are signs in exactly the same sense of the word "sign" that black clouds are a sign of rain . . .' (1971, p. 429). If he is correct, then the 'semantic' status of bee dance, if we suppose the bruise to lack such status, is to be recovered from certain features of the interactions between senders and receivers of the messages in question.

Indeed, this is a challenge that the materialist should be willing to accept. He can analyse the distinctive features of communicative interactions while still using the language of the mentalist. That is, he can attribute beliefs, desires, intentions, and so forth, in giving a simpler account of a complex matter. The complexity of the account will increase vastly when replaced by ultimate 'physical' talk, if this indeed be possible. Nevertheless, the complexity of the faithful physicalist version should not blind us to the features that distinguish the causal processes in the bees' brains from those involved in the swelling after the swipe. The difficulty lies precisely in spelling out what these distinctive features are, and how they are pertinent to the contrast between conveying information, on the one hand, and, on the other, being black and blue.

So what is there to linguistic meaning, over and above mere causation? One might offer the following reason why one could regard bee dance as having communicative significance lacking in the case of the bruise: that the dance is a somehow 'arbitrary' or 'rule-governed' causal product of the sighting of food. It is arbitrary in that the rules (consciously followed or not) in accordance with which this behaviour is produced, *could have been otherwise*, while yet serving the same communicative purpose. That is, as a piece of adaptive behaviour, whether wholly instinctual or partly learned, it may very well *now* follow as a causal consequence of the sighting; but that precisely *this* sort of dance should have come to serve this purpose is, in a phylogenetic perspective, quite accidental. Selective pressures for a system of communication of this kind might have produced, through different mutations for new behaviour, a different dance or indeed wholly different behaviour types, serving the same function. In fact, another species of bee that communicates about food sources (*Apis florica*) does so in the same way except in so far as it wiggles in the horizontal rather than the vertical plane, and so can indicate directly the angle of the food source from the sun. *Apis mellifera* (the common bee) has to have a 'convention' as to which angle in the vertical plane corresponds to this latter angle. *Apis mellifera* if put on a horizontal honeycomb, dances just like *Apis florica*. (This 'conventional' behaviour is described with scare quotes because it is not clear that one would be justified in reading into it the constellation of reciprocal beliefs and intentions that someone like Lewis (1969) takes as constitutive of conventional behaviour.)

Similar considerations apply to other systems of animal communication, especially those more rudimentary than bee dance. The dance of the bees is a *behavioural* system. But it is quite possible to find cases in which a fixed physiological feature functions as a signal. For example, the red spot on a female herring gull's beak is a signal to her young that food is being presented. The spot serves an adaptive purpose and has evolved with it in the same way that bee dance has evolved with its purpose. In this sense, then, the spot is a signal – the single one in a very rudimentary system.

The spotted beak is an example intermediate between the bruise and the bees' dance. A system of signalling similar to the bees', but in which the signals are learned and not genetically pre-programmed, is that of bird song in certain species. The same type of bird song can serve different purposes, depending on the context in which it is emitted. Such purposes could be as varied as warning off male intruders, attracting female mates, etc. Thus the 'meaning' of the signal depends on the sex of the receivers. It could also depend on the time of day, the motivational state of the receiver etc.

So the suggestion before us is that communicative significance derives from the phylogenetically arbitrary status of the evolved form of behaviour. It is but one among many possible solutions to the species' co-ordination problem. But the suggestion is open to an objection, which we owe to David Hull. This is that (in the case of the bees, for example) the dance might have been the only way to solve that problem, given bee physiology immediately before its evolutionary debut.

Now although in evolutionary terms, given the amounts of genetic variability at hand, it is likely that such behaviour has been arbitrary in the required sense (witness *Apis mellifera* v *Apis florica*), one must concede the conceptual point here. Armstrong even goes as far as asking 'is it not conceivable that the whole of syntax and semantics should have been innate so that all mankind spoke the one, wired-in, non-conventional language?' (loc. cit., p. 437)

What, then, would justify the attribution of content to exercises of evolved forms of behaviour? A promising line of thought due to McDowell (1980) is that all one requires is *transfer of information* simpliciter. This holds regardless of whether the mechanism, behavioural, chemical or otherwise, might have evolved arbitrarily or might have been the only possible one phylogenetically. The natural function of this communicative behaviour, whether wholly instinctual or not, is to impart to fellow creatures 'cognitive stand-ins for the states of affairs they represent'.

> The function of such behaviour is to furnish information about the environment to (bees) which witness it; . . . Aims pursued in communicating do not enter the story. There are no such aims since the behaviour is instinctive. . .
>
> When the communicative process functions properly, sensory confrontation with a piece of communicative behaviour has the same impact on the cognitive state of a perceiver as sensory confrontation with the state of affairs which the behaviour, as we may say, represents; . . . elements of the communicative repertoire represent states of affairs by virtue of standing in for them in a creature's cognitive dealings with the world.

Compare now the case of the bruise. Bruising, it is reasonable to suppose, has never served the adaptive function of informing others that the body has been hurt. As a consequence of the blow it has wholly to do with internal processes of tissue regeneration. So far, so good; we have separated bee dance from bruises. What about us? Is not a belief/intention ingredient essential to the notion of linguistic communication in the human case?

3 BELIEF, INTENTION AND MEANING: RADICAL INTERPRETATION

Certainly the Gricean analysis (1969) would have it so. On this highly influential account of linguistic meaning, certain 'higher order' intentions are crucially involved. In the case of assertion, for example, if one utters x, thereby meaning that p, one does so intending

(i) that the audience should believe p
(ii) that they should be aware of this intention, and
(iii) that this awareness should be part of their reason for believing that p.

The question that McDowell is pointing to, however, is whether, in this analysis of human communication, Grice's contribution concerns more what it is to be human than what it is to communicate.

Whether a system with finitely many basic signals is a linguistic system depends on whether the characteristic behaviour within which and against which it is interpreted sustains attributions of higher order beliefs and intentions to the creatures concerned. If one has to resort to such an intentional framework in order to make sense of the behaviour, in order to explain what the creatures are doing and why, and has so to resort after exhausting all possibilities of more austere, less anthropomorphic frameworks of explanation, then one is on the route to linguistic meanings. Wittgenstein (1953) invites one to consider a primitive signalling system as the system of communication of a whole tribe. We can acquiesce in this thought experiment only because we implicitly assume that we shall discover behaviour forcing us to regard tribe members as having beliefs and intentions, even if, *ex hypothesi*, their very simple language does not permit them to express these beliefs and intentions. Provided a tribe member can recognise an intention that he should respond appropriately to a signal in the simple system, we are happy to regard the signal as linguistically meaningful to him, rather than simply causing or triggering certain behaviour – as was the case with the bees.

When radically interpreting the speech of a totally alien tribe (or species), one has to consider all the evidence there is concerning mental states. Their attributions go hand in hand with, and must be adjusted to, our attributions of meaning to utterances. For example, a communicative act is a lie if by ascribing the appropriate belief and intention to produce in the audience the opposite belief, one achieves the best explanation – within one's theory of the agent's beliefs, desires, etc. – of his act.

Despite the importance of the intentional aspect of language in the comparison of human with other animal communication, we must not lose sight of the important role to be played by other considerations. These include range of vocabulary items, methods of syntactic combination, discreteness of signals, and the associated semantic possibilities. Among the latter are reference to spatio-temporally remote items, generality and tense – to name but a few. They are among the most important semantic features of human language for which one would understandably be hard put to find correlates in other known systems of animal communication. There seems to be no evidence so far that non-primate signalling systems enable reference to objects or events in the remote past or future.

A past event can influence aspects of a signal without the latter being a signal about the past. We need evidence of a systematic frame of temporal reference – involving analogues of 'now', 'then', 'soon', 'earlier than', etc. – before accepting a specific tensed interpretation of any signal in the system. This point would certainly be reinforced if one held that if we were prepared to interpret an animal's signal as one to the effect that p, then we should also be prepared to attribute to the animal the belief that p. This granted, we would have to agree with Bennett when he writes: 'No doubt past beliefs are also caused by past experience; but what I am denying is the converse claim that any belief which is caused by the past must be or involve a belief about the past' (Bennett, 1976, p. 103).

Closely related to the objection that animal signallers cannot 'say' anything significantly tensed, is the objection that they cannot 'say' that all F's are G's, or that all F-events are followed by G-events. One has to distinguish between a general belief's being simply causally responsible for the emission of a signal, and the signal itself being an expression of the general belief. It is of the essence of the linguistic expression of a generality (All F's are G's) that it be not tied, for its meaning or point or appropriateness, to any specific F-stimulus or set of F-circumstances. The creature should be able to convey this general information to its fellows without the presence of an F to prompt him otherwise his signal could not mean 'All F's are G's.'

Considerations concerning syntactic and semantic structure are not unconnected with those about the beliefs and intentions involved in linguistic communication – beliefs and intentions which McDowell suggests are quite secondary once one has accepted that communication in general (from simple instinctive types through to natural language) has first and foremost to do with the transfer of information, or the instilling of knowledge.

We wish to suggest that other considerations of an evolutionary kind lend weight to the view McDowell is challenging. To do so we must sketch the bare outlines of the opposing views on the conceptual issues, to see which one harmonises more with evolutionary conjectures about language.

4 TWO EVOLUTIONARY ACCOUNTS

The first evolutionary account, which places more emphasis on the intentional ingredient than McDowell would allow, is as follows.

Following Bennett (1976), we regard it as admissible to frame theories of perception (or 'registration') and of 'proto'-belief and desire to explain the actions of certain creatures. But, importantly, the theory allows attribution of belief and intention *at higher order*, even in the absence of language. Then, when higher order beliefs and intentions of interactants engage in a suitable way – by instantiating the Gricean mechanism, say – certain actions (token events) can have *occasion meaning*. This is meaning in as rich an intentional sense as one is likely to get – far more than the mere instilling of (first order) knowledge with which McDowell deals. Now, says, Bennett, we can move from meaningful event tokens to conventional meanings of event types. For we are dealing, *ex hypothesi*, with creatures who can grasp the salience of certain features of past occasions of meaningful exchange. Lewis's (1969) theory of convention is then applied to account for the acquisition of conventional meaning by certain action (or utterance) types in the developing linguistic life of the group.

Communicative doings are governed by social conventions which ensure that the hearer (rightly or wrongly) diagnoses intentions and forms beliefs in the way Grice requires. If a supposed act of communication does not involve stock signals or familiar moves (albeit in new combinations), but is unprecedented and *ad hoc*, then it can be very difficult for the audience to discern from the available behavioural evidence that Gricean intentions are present.

Yet it is just the presence of such intentions that it must have been possible to diagnose *ab initio*, in the case of so-called 'occasion meaning', before the growth of linguistic conventions that now enable ready diagnosis and indeed make it virtually superfluous.

When a particular primitive utterance type begins to be associated with certain salient features of the situation, and is used again and again to similar effect as before when those features recur, we can witness the growth of a *convention* to the effect that that particular utterance type will henceforth mean that p. But it is a convention that is established not by convening, as David Lewis has made engagingly clear, but rather one that grows by accretion and solidification of behavioural forms, the pattern becoming entrenched as a stable form of social interaction within a network of mutual and reciprocal beliefs and intentions and expectations. In this way the earliest undifferentiated forms of some proto-language might have become endowed with conventional meanings. It is at this point that we might have to accept the possibility of 'thought without talk', *pace* Davidson (1975), to account for how language might have evolved.

But what is of crucial importance in the account thus far is that these action types are syntactically and semantically *unstructured*. At this point in Bennett's exposition there is an interesting leap (between chapters 7 and 8). Having accounted for how action types might acquire conventional but unstructured meanings, he advances straightaway to a discussion of how one might come into an alien community and find evidence that their linguistic interactions are *structured* (syntactically and semantically). Yet he offers no account of how the language of the group could have attained such complexity of structure from the humble (albeit highly intentional) beginnings he had been discussing earlier. Let us call this the grammar-gap of the intentional account.

Compare now McDowell's account, which suggest a second scenario.

> A more attractive line of thought is that the linguistic repertoire retains, through the alteration of nature involved in the onset of self-consciousness, a form of the characteristic which was essential to its pre-linguistic ancestor: in suitable circumstances (to be spelled out in any fuller elaboration of this idea) its exercises are cognitive stand-ins for the states of affairs which they represent. (*loc. cit.*, p. 135)

This opens the possibility that communication could involve *structured* messages, perhaps even structure of the sort producing infinite generative capacity, without the creatures' ever having passed the 'Gricean hump' that Bennett put them past even before occasion meaning had made its evolutionary debut.

It is worth pausing here for a moment to reflect on the weight of traditional thinking that McDowell is seeking to shift:

> . . . the utterance '(There is yellow fever on board' . . . is a sign that (the speaker has the objective that an audience should have reason to believe (that the speaker believes [that (there is yellow fever on board)])) . . . (Armstrong, 1971, pp. 432-3, 435)

> . . . in order to make an assertion a speaker must represent himself as believing what he says. (Davidson, 1979, p. 12)

> in the performance of each illocutionary act with a propositional content, we express a certain Intentional state with that propositional content, and that Intentional state is the sincerity condition of that type of speech act. Thus, for example, if I make the statement that p, I express a belief that p . . . (Searle, 1979b, pp. 75, 78)

What McDowell is challenging is the notion thus unanimously upheld that the belief-expressive feature of assertions is essential to our understanding of the extra-linguistic purpose of statements of a natural language: challenging it, indeed, even for statements with syntactic structure.

What we wish to suggest is that filling the grammar-gap makes structured talk without higher order thought look much less likely. Even if one were to concede to McDowell that there could be traffic in simple message types before the Gricean hump, we wish to maintain that one would have to be over the hump (or at least capable of being in such states as are involved in the hump) before one's language could evolve syntactic structure of the kind yielding infinite generative capacity. We stress 'of the kind yielding infinite generative capacity' because it is the peculiar combinatorial potency of expressions in natural language which, for reasons given below, could hardly have emerged by any evolutionary mechanism unless the users of the system were capable of the sort of second order intentional states involved in the Gricean mechanism. We are quite willing to countenance the possibility that structure might usefully be discerned in signals whose segments thus descried were iteratively impotent. Different temporal segments of a mating display, say, might convey information about different aspects of the physical prowess and motivational states of the performer. Likewise different simultaneous aspects of a dance, such as wiggle and tilt with the bees, might separately convey information about different aspects of the world (distance, direction and quality of food source). But expressions of a recursive language, we maintain, could probably not have acquired their categorial valency as it were, their powers of combination into new but readily understood messages, unless their users were on the second rung of the intentional ladder and therefore, potentially, past the Gricean hump.

To make this claim plausible – a claim which, it seems to us, lies in a grey zone between the conceptual and the empirical – we wish first to offer what seems to us the most likely account of the evolution of syntactic structure. Even if our assessment of its implications concerning the relative order of emergence of the intentional ingredient and of syntactic structure were held to be incorrect, the mechanism of that evolution might be of independent interest, and be seen as bearing on other problems besides (especially in developmental psychology and theoretical linguistics).

5 THE CHOMSKIAN PARADIGM

First, however, we must briefly explain the details of the Chomskian paradigm in the grammar of natural language. No one is better known than Chomsky for his insistence that language is a special cognitive capacity innate in man and unique to him as a species. In recent writings he has put forward a picture of language growing as a mental organ, not being learned, but simply unfolding or blossoming as the genetically encoded open programmes process linguistic stimuli. One's 'knowledge' of language – or at least knowledge of that aspect of it which linguists wish to capture in their systems of grammatical rules and representations – is therefore *implicit* knowledge, of a kind not readily articulable by one who possesses it. Perhaps the most important reason for Chomsky's botanical model of language acquisition is the following. Any human infant placed at birth in any linguistic community will come to speak their language as fluently as the natives; and indeed will do so with remarkable rapidity, mastering a very complex system of rules on the basis of scanty data. The scanty data is, of course, the sum total of the child's linguistic experience, including not only the utterances to which he has been privy, infelicities and all, but also the observable reactions of fellow speakers when he himself attempts to speak.

It is contentious whether the available data severely underdetermine the notion of grammaticality attainable by one who is not programmed to select a grammar from an already narrowly constrained class, but who must instead cast around among all logically possible conjectures as to how the sentences of the language might be enumerated – or, better still, as to how to decide in general whether a given utterance is a sentence of the language. Words like 'scanty' and 'rich', however, as applied here to data bases that are to support a 'theory' of grammaticality, are notoriously difficult to make precise. Their numerical quantification is almost out of the question. But those in broad agreement with Chomsky on the qualitative point he is making will point to the myriad possible sentences that even a child speaker could produce or understand without ever having been exposed to them beforehand. Certainly, it does seem as though the child must have mastered, albeit implicitly, a recursive means of generating or recognising sentences of his language, and working out their meanings. Otherwise it is difficult to see how this potentially infinite capacity could develop on the basis of finite experience. The major question then would be – are these recursive resources an innate, species-specific endowment? Moreover, are they a *task*-specific endowment? Do we have resources specialised just for the acquisition of language? Is Chomsky really saying more than that any learned behaviour must have a genetic basis that makes learning possible? Without providing any specific hypotheses about cerebral localisation or compartmentalisation of function, Chomsky believes that the brain – innately, and perhaps also after maturation – embodies the 'modules' of a transformational grammar. Initially the neural wiring is such as to dispose the learner to home in on rules of a certain kind – the kind, namely, that make up a Chomskian grammar. The initial wiring is our species' common genetic heritage. Experience of the local tongue then sets certain values for previously unspecified parameters, and the grammar starts taking shape.

> Each end state will be one of the systems compatible with the system of principles, but there is no reason to expect that the resulting end state is derived inductively (or by any general procedure) from the data that trigger and shape language development. In a roughly comparable way, embryological development involves the unfolding of a genetically determined program to an end state that is not derived by any general 'growth principle' from the environmental inputs, and slight changes in the program – say, in the timing of regulatory mechanisms – can lead to radically different organisms. (Chomsky and Caplan (1980), p. 98)

What Chomsky (and his co-author) are here saying is that the acquisition of language follows a very open genetic program. The idea is old, the style of presentation newly Chomskian. What, then, is the rich system of principles in question? A Chomskian grammar, in outline, consists of a base of phrase structure rules, a transformational component and a phonological component. The base generates so-called deep structures by various applications of the phrase structure rules in various orders. Deep structures wear their basal pedigree on their sleeves. They are complex abstract objects consisting of grammatical labels arranged in a tree pattern, such as

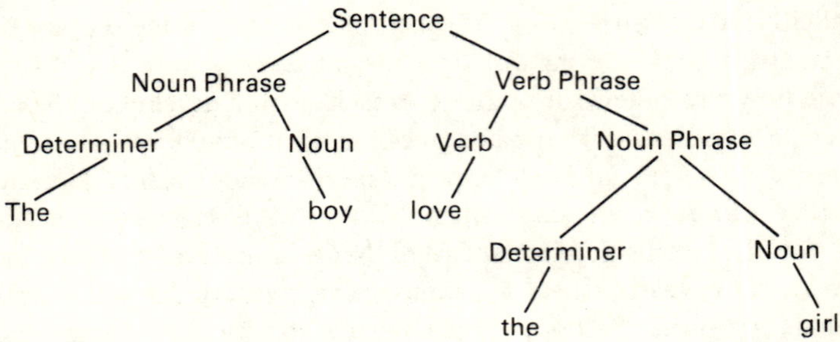

Figure IV.1

In this example, which is, of course, very simple and chosen only for purposes of illustration, one can clearly see the natural breaks that we intuit in the structure of the sentence. These are represented by the branching pattern of grammatical labels. The first branching occurs according to the phrase structure (or re-write) rule

$S \rightarrow NP + VP$

Then we apply the rules

$NP \rightarrow Det + N$
$VP \rightarrow V + NP$

and finally, at bottom right, the rule

$NP \rightarrow Det + N$

once more. Then we have performed certain lexical insertions, observing the categories of lexical items called for by the terminal labels. In our example the homely English sentence 'The boy loves the girl' looks as though it can now be read off straightforwardly from the deep structure. But, of course, we have been simplifying matters greatly; ignoring the question of subject-verb agreement as registered in the tell-tale 's' of 'loves'. In an inflected language such as Latin, such matters will be altogether less tractable, even in the case of a similarly terse sentence. This is where the transformational component comes in. Its function is to take such a deep structure and successively transform it into the completed 'surface' structure, with words correctly ordered in linear sequence, correctly inflected and so on, marking the whole in such a way that the phonological component can then render it as the appropriate sound pattern. It was the transformations that transformed linguistics when Chomsky first published his seminal book *Syntactic Structures* in 1957. It is the transformations that turn actives into passives, bring material to the beginning of the sentence for peculiar emphasis, ensure that pronouns much later on in the sentence pick up antecedents at the very beginning, and so on. Chomsky argued forcefully that other rule schemes available at the time – such as, indeed, a scheme consisting exclusively of phrase structure rules – were inadequate to the task of generating the grammatical sentences of a language like English, and of endowing them in the process with the grammatical structure that speakers intuitively regarded them as having. The positive side of the Chomskian paradigm has since been to elaborate the rules that will do the generating. Very importantly, the transformations were to be subject to *constraints* that would prevent them from achieving the desired effect by 'brute force', thereby trivialising the problem of describing a complex but orderly phenomenon. Various constraints were proposed, to do with the ordering and scope of transformations, and whether they were to be applied obligatorily at certain stages or not. The linguistic literature burgeoned with specific proposals and then flocks of counter-examples. Linguists concentrating on the local behaviour of a certain kind of adverb, or system of verb endings, say, would find their proposed rules succumbing to examples whose peculiar intractability lay in other parts of speech so importantly related to the behaviour of those adverbs or verb endings that the latter could not be treated independently of them. It was as though any thread in the whole theoretical tapestry could be pulled in such a way as to run ladders through the whole system.

According to Chomsky's model, however – such problems of detail being set aside for the present – the brain is equipped with several linguistic universals in the form of general rule schemes or constraints governing transformations. Every natural language is thus cast in the same general mould; but its detailed pattern is not so much learned sentence-by-sentence, but rather pinned down by particular choices of values for important parameters.

Let us briefly summarise the nativist views propounded by Chomsky.

1 All natural languages display syntactic 'universals', which can be detected and formulated within the theoretical framework of a Chomskian grammar.
2 These grammars are to be preferred to all their competitors on grounds of the latters' demonstrably inadequate generative power.
3 Because children learn the 'correct' grammar so quickly, with such limited exposure to grammatical sentences, they must have an inborn capacity to construct the right kind of grammar – that is, the kind of grammar displaying the observed universals.

6 A NEW MODEL FOR THE EVOLUTION OF LINGUISTIC STRUCTURE

Recently Sampson (1978) has challenged the status of many of the alleged linguistic 'universals' put forward by Chomskians. The main syntactic universal upon whose existence both Chomsky and Sampson are agreed is that all sentences of natural language are *hierarchically structured*. It is with respect to this universal that we can illustrate the novelty of Sampson's contribution. He has argued that 'independent explanation, more plausible *a priori* than nativism, is available for the universality of the trait'. His attack is therefore two-pronged: to argue that there are far fewer genuine universals than had been thought, and to explain away the remaining one in a way more plausible than Chomsky's.

Any sentence can be broken down into immediate sub-units which themselves have reasonably independent status as coherent wholes, whose combination produces the original sentence. Each of the sub-units may in turn be decomposable into sub-sub-units and so on, until non-decomposable words or morphemes are encountered. Thus the revealed hierarchy of expressions is a tree-like structure with branchings determined in a definite way. It was precisely this analysis of how certain expressions lower down in the tree thus fell within the scope of others higher up on a branch that launched modern logic on its powerful and sophisticated account of logical relations among sentences.

Chomsky and Sampson claim that this tree-like decomposition of a sentence into its parts is a feature common to all human language, and is indeed the basis upon which we determine the meaning of a sentence from the way it is built up out of its parts (an insight going back at least to Frege (1879)). In a Chomskian grammar it is the base component, consisting of phrase structure rules, that generates the sort of revealed structure just explained. According to Chomsky, we are genetically pre-programmed to search for just this sort of structure underlying the sentences that we hear as infants. The strict sense in which such structures underlie the sentences is of course enshrined in the transformational component of the grammar. All this imposes a powerful constraint on language acquisition from the allegedly scanty data available to any child. The alleged deficiency is compensated for by a neurologically based preference for *Chomskian* grammars. That is, the child is pre-programmed to arrive at *this* sort of representation of his mother tongue rather than any of the many other mathematically possible kinds. The representation is, of course, implicit and we would not expect any child to be able to state explicitly the phrase structure rules and transformations generating the sentences of his language. There is also considerable evidence (as marshalled by Ingram (1975)) to suggest that up to the age of six years it is mainly short, simple but still hierarchically structured sentences that the child is able to produce. Transformations that have to do with embedding one sentence into another (such as subordinate clause or relative clause formation) are only acquired between the ages of six and twelve. This would imply at least that the early fragment of language poses the problem of hierarchical analysis in a reasonably pure form.

Sampson's attack on Chomsky's account of how the child overcomes this problem is ingenious. He advances a quite different explanation of why hierarchicality is a feature to be found in all languages. In so doing, he avoids any appeal to neurological pre-programming of a specifically linguistic kind as claimed by Chomsky.

He suggests an evolutionary model due to Simon (1962), originally devised for quite different domains than the present one. These include the evolution of human institutions, and various kinds of pre-biotic evolution. The idea is strikingly simple. If one considers hierarchical assemblies in general, their evolution from their constituents is far more likely if they consist of relatively stable sub-assemblies which themselves are evolutionary products of an earlier period. Evolution proceeds by the accidental combination of already existing stable sub-assemblies, thereby producing new stable assemblies of higher complexity. A new stable assembly establishes its credentials as a useful whole, and is co-opted into the ever-growing network of items with burgeoning structure. If we have a principled way of discerning the stable sub-assemblies, thereby distinguishing them from merely arbitrary collections of parts, then we can as it were see the evolutionary pedigree of a complex structure.

Let us apply this account to the syntactic structures of a natural language undergoing the process of evolution which it is only reasonable to suppose has taken place. The systems of communication of our primate ancestors presumably consisted of words and short sentences (but what we have to say would hold even if only gestural sequences were involved). It is reasonable to suppose that these language users might (even accidentally) have hit on new combinations of phrases to produce slightly longer sentences than had hitherto been the rule: sentences, moreover, whose newly coined significance derived from both the context of their first use and the pre-established significance of their components.

Now there are two ways that the new composite sentence token will be of a type that is eventually to acquire a constant significance. (Note here that we are doing our best to avoid speaking of conventional meaning, for we do not wish to beg the question yet against the second scenario.) The first is a highly fortuitous and austere way: namely, that the compound might have the same effect – the transfer of some particular kind of information – sufficiently often and sufficiently advantageously for there to be some selective advantage in its coming standardly to possess that significance: that is, for it to acquire the biological *function* of transmitting just that kind of information.

In the case of instinctual communication the fortuitous nature of this process is only too apparent; for it requires the formation of new closed genetic programmes for both the production and processing of the complex signal. Another observation is crucial here. Remember we are concerned eventually to account for the repeatable contribution constituents make to the significance of signals involving them; this being of the essence when syntax is recursive, or creative. Thus we want to see the representational role of the newly juxtaposed elements as recognisably preserved within and thereby helping to determine the new representational role of, the new signal in which they occur. (We use the phrase 'representational role' rather than 'meaning' here to hew as closely as possible to the line of the second scenario.) Now in a language of potentially unlimited generative capacity, we fail to see how this could be achieved for all expressions across all patterns of combination in the first way described above. In so far as qualitative argument about improbability of emergence could have any force, surely this is one such case.

Compare now the vastly more plausible second way in which the new composite sentence token may be of a type that is to acquire constant significance. And let us no longer worry about begging questions against the second scenario, and simply say 'conventional meaning' rather than 'constant significance'. The second way makes the emergence of syntactic combinations seem much less fortuitous.

We take up an intriguing claim by Schmitt (1955). A proto-language might contain one-word sentences used appropriately (in a variety of speech acts) for things and events. There might be words for 'man', 'seal', 'booty', etc. and perhaps also words for events such as killings or harpoonings. Bearing in mind constantly the radical interpreter at one's shoulder, one could regard these as mainly nominal in character. Facial gesture and general demeanour could easily signal the force with which the utterance of such nouns was to be taken. Truculent and threatening utterances would be demands, plaintive ones requests, excited but friendly ones declaratives, and so on. Schmitt's claim is that even in present-day Eskimo the verb translated into English as 'kill' is a peripheral modification of the noun for 'booty'. One can imagine the first fortuitous utterance of the three-word string

Man seal booty

as having the occasion meaning (via suitably diagnosed Gricean intentions, etc.) of something roughly like 'The man killed the seal.' New strings like this, in a highly salient context, could readily be understood as new messages, of a new level of complexity, by an audience who already grasped the components from their isolated usages in the past. We may expect new conventions governing syntactic combinations – in our example, the subject-object-verb complex – to establish themselves quickly in the evolving language of any group whose members are bright enough to tumble to the meanings of such innovations. But this crucially involves their being past the Gricean hump.

As new syntactic categories settle down by innovative accretion along the lines suggested, the complex mesh of grammatical relations becomes the cloth of an ever-changing community coat. By the time transformational rules (if they are really operating – which is theoretically contentious, as we shall see in due course) enter the picture, we can expect even more linguistic material to be available for reshaping, re-ordering and relocating. In this way the surface output, being as it is so critically dependent on transformational pedigree, becomes a highly labile and volatile product.

Our account of innovative accretion tells us how the earliest linguistic structures were born. A few salient juxtapositions would confer upon words a new relational potency over and above their semantic directedness to the external world. Doubtless that semantic directedness is an indispensable clue, for later inheritors of the developing language, as to how sentences are put together grammatically. Indeed, Pinker (1979) has pointed out the considerable advantages, in mathematical models of language acquisition, in having such semantic clues. The language acquisition device, in its attempt to unravel prevailing grammatical structure, would therefore do well to pay special heed to the sorts of semantic contexts which (we are suggesting) were the evolutionary source of those very structures. Used in new combinations to salient contextual effect, words would acquire a certain potential for juxtapositions, a grammatical valency, which is now captured by the idea of the *category* of an expression, due to Frege (1879) and Ajdukiewicz (1935). Phrase structure rules would implicitly have been adopted, legitimating previous combinations that have successfully occurred, as well as others that had not yet occurred but might very well have, and with similar success, had the choice of words and message had to be different. Thus even the earliest and most modest collection of phrase structure rules would have been pregnant with new output; and as the rules became entrenched so too would innovative effort and diagnostic insight become more relaxed and automatic, as words appeared newly combined in well-understood syntactic contexts. Transformational rules also, presumably, emerged at this time – at first, perhaps, with mainly abbreviatory effect. Later they could have wedded sentences in constructions such as causal conditionals, being both a spur to and the reflection of our ancestors' growing conceptual mastery of the world about them. Premack's tests (1976) to establish that chimpanzees make rudimentary connections between 'states of affairs' (in order not to beg the more refined question of objectuality within them) which to us appear closely connected by virtue of a cause and an effect, or by virtue of an implicit goal or problem and a means or stratagem, seem to support attributions to chimps of protean correlates of human categories of thought. Indeed some primatologists and psychologists seek to apply Piaget's theory of stages of cognitive development to make comparisons between man and other primates. They propose tests for mastery of such notions as object-permanence and conservation, and apply a rich fund of observations to rank baboons, chimpanzees, gorillas, orang utans and various monkey species in respect of their dexterity and inquisitiveness. (See Braggio *et al.*, 1982; Thomas, 1982; Vauclair 1982.) Wynn (1981) has even gone so far as to suggest a position on Piaget's scale for an extinct hominid ancestor in the Olduvai gorge in East Africa! This would involve judging the noûs needed to fashion the sorts of tools whose remains have been discovered. Parker and Gibson (1979) and Lamendella (1976) have taken Haeckel's principle more or less at face value in the phylogeny of cognition. They suggest that human infants' growing conceptual mastery of the world about them recapitulates the evolutionary order in which our primate forebears attained it. Thus the child's linguistic beginning would be on a rough par with the protolanguage of early hominids. It is not too fanciful to see a connection here with a point in Piagetian spirit that Ingram (1975) has made about the acquisition of linguistic transformations by children (to dwell within the Chomskian paradigm a little longer). It invites one to see Haeckel's principle at work once more, with ontogeny recapitulating phylogeny:

> most of the transformations of English are only acquired between the ages of six and twelve. These are the transformations that have to do with the embedding of one sentence into another. The evidence for this comes from both theory and facts. The facts come from the observation of children's complex sentences between two and twelve. The structures up to four are primarily simple sentences. Those from four to six show complex structures, but most propositions simply juxtaposed to each other. Piaget's theory of cognitive development predicts that this should be the case. To relate two structures to each other, the child needs to be capable of performing concrete operations. This ability is only developed between the age of six and twelve.

As Sampson observes, the phylogenetic account above of the simultaneous growth of phrase structure and transformational rules yields a prediction agreeing nicely with a principle formulated by the Chomskians. This principle states that transformations apply cyclically, and that later transformations, in tinkering with the transforms already produced, deal with them as completed units undergoing at most peripheral changes in this later re-arrangement. That is, later transformations tend not to interfere with or undo the effects of earlier ones in the generation of a surface from a deep structure. This is precisely what one would expect if the above evolutionary account were true. One would expect existing sentence patterns to be stabilised in use, and to be only minimally disturbed upon being combined to form more complex syntactic units. According to Sampson, many other features of transformational grammar can be explained away in a similar fashion once one adopts this evolutionary perspective. We would point out in this connection that two other constraints, originally proposed to ensure the learnability of transformational grammars obeying them, are highly likely corollaries of Sampson's account.

These principles are, in Pinker's words (1979), the 'Binary Principle', to the effect that 'no transformation may have a structural description that refers to symbols in more than two adjacent levels in the deep structure', and the 'Freezing Principle, which forbids a transformation to apply to a configuration of symbols that could only have been created by the previous application of another transformation'.

Although Sampson himself does not point to independent evidence or hypotheses about syntactic change, what he suggest coheres well with the speculations of Chung (1977). From evidence concerning Pukapukan and Samoan, she forms the following conjectures. Syntactic change affects simple sentences before it affects the action of superficial rules; and when transformational rules are affected, the more superficial ones are affected first, the major cyclic ones last.

The grammar-gap in the intentional account is, we are suggesting, to be filled by something like Sampson's account of the evolution of hierarchical arrangements of stable sub-assemblies. Innovative combinations become part of a familiar repertoire precisely because they offer a form of solution to recurring co-ordination problems whose salience the system-users are quick to discern. And the accompanying growth and refinement of categorial conventions is possible only because the system-users have higher order beliefs and intentions.

A further suggestion that flows from this is that such evolutionary considerations lend support to a competing grammatical paradigm – that of Montague grammar (Montague, 1974). Whereas in Chomskian grammar the basic approach is 'top down', with transformation rules sometimes applying in ways that require one to consider syntactic environments beyond the immediate focus of application, the picture in Montague grammar is simpler, in a way more congenial to our evolutionary picture. In Montague grammar various categories of expression are simultaneously defined by an inductive process that builds up expressions across the whole language. One may think of similarly structured expressions within one category – differing only in lexical items – as stable sub-assemblies with an independent communicative value of their own. This may derive not only from repeated occurrence within wider syntactical contexts, but also from frequent usage 'in isolation', as it were. One thinks here, for example, of noun phrases being used in response to wh-questions. (Indeed, one empirical reason why a certain category within a Montague grammar should be basic may be precisely a high pragmatic probability of its isolated use in fruitful exchanges: here the grammar of conversation rather than of the sentence may become crucial. One is inclined to ask here just how well-argued is the frequent claim that 'the *unit of communication* is the sentence' (Armstrong, 1971, p. 428).)

The way a sentence (or indeed any complex expression) is generated within a Montague grammar provides a possible synchronic fossil of how, diachronically, the language acquired the complexity making such a sentence possible. For in Montague grammar one begins with lexical items (of known categories) and 'assembles' lower level structures. Within these one can still discern the 'earlier' items, whose syntactic combination in accordance with Montague's formation rules involves only relatively minor peripheral modifications (just what one would expect on the Sampson-Simon model). The new lower-level structures are then themselves syntactically combined, again with peripheral modifications not obliterating their essential unity, into yet higher ones; and so on, until the sentence in question is produced. Importantly, there are no transformations capable of wholesale disfiguration of the by-products of any stage of the generative process.

Montague generation, proceeding as it does through all categories of expression simultaneously, promises recognition of the conversational integrity of parts of speech in a way that sentence-focussed Chomskian grammar does not. Moreover, there is a further independent reason for finding it attractive, apart from whatever success is to be had in the future in generating likely looking fragments of natural language. This reason is that it is mathematically much less powerful than Chomskian grammar. Peters and Ritchie (1971) have shown that any language all of whose sentences can be mechanically listed can be generated by some Chomskian grammar. Montague grammars, by contrast, characterise at most context-sensitive languages (see below), and therefore yield decision procedures for grammaticality.

We may indicate the way in which this sentence has just been constructed by means of the following *analysis* tree:

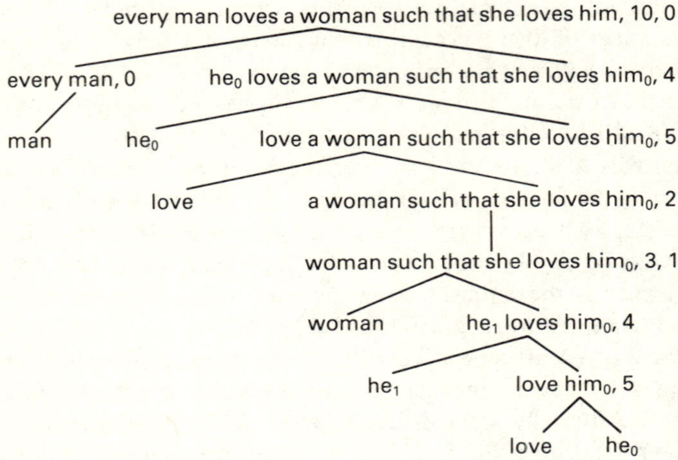

Figure IV.2

7 PROBLEMS IN PARSING

This is especially desirable in the light of our manifest ability to parse – that is, to produce judgments as to grammaticality, and not just judgments that are grammatical. It also means that whatever part of the brain it is that deals with matters grammatical can be regarded as relatively low in the so-called sub-recursive hierarchy of computing machines – putting it within easier reach, perhaps, of the evolution of cognitive capacities 'from below'.

We wish now to venture upon an area of some technicality that bears most interestingly on this sort of evolutionary consideration. A Turing machine is a mathematical idealisation, intended as a model of a so-called *computable* function. Turing machines are the most general kind of computing device possible. For, according to the so-called Turing-Church thesis, *any* computable function can be computed by an appropriately chosen Turing machine.

A Turing machine is basically a 'black box' capable of at most finitely many states, equipped with a tape that is infinite in both directions. The tape is marked off into cells. The machine can scan one cell at a time. Each cell will be empty, or will have a 'one' written on it. As the machine scans a cell, it performs a certain action and enters a new state. What actions it performs, and what state it enters, depends only on the state of the scanned cell, and the present internal state of the machine. The possible actions are:

(i) erase a 'one'
(ii) insert a 'one'
(iii) move one cell to the right
(iv) move one cell to the left
(v) halt

Thus if we feed a sequence of n 'ones' into the machine in its initial state, what it does is uniquely determined by its 'programme' – that set of instructions that tell it what to do, and what state to enter, given its present state and the state of the presently scanned cell. The programme could, for example, be that of the doubling function. In this case the machine will process the n 'ones' we feed into it, and eventually have written 2n 'ones' on the output side of the tape before halting.

Any algorithmic transformation on finite structured objects can be effected by means of a Turing machine. For the objects can be *coded* as numbers. The machine can be fed a code number in order to compute another number from it. The latter can then be *decoded* to find the 'target' structured object resulting from the transformation in question. Thus, for example, transformations on phrase markers could be treated as Turing computable functions.

Now a *grammar* for a language is simply a method of enumerating or listing all and only the grammatical sentences of the language. An enumerative grammar might, then, be represented by a computing machine. Given the number 1, the machine calculates the code number of the first grammatical sentence listed by the grammar. Given the number 2, it calculates the code number of the second grammatical sentence; and so on.

The objection to Chomsky that now emerges is the following. Merely requiring that a grammar should mechanically enumerate exactly the grammatical sentences entails (via the Peters-Ritchie result mentioned above) that, of necessity, some Chomskian grammar will accomplish the task. That human languages have Chomskian grammars is therefore quite uninformative. The so-called 'transformational constraints' within these grammars prove to be no constraint at all on what the grammar can produce.

So consider now any Turing 'black box' that accomplishes what a Chomskian grammar does, but by very different computational procedures and internal wiring. How are we to choose between the two? Perhaps we can invoke the extra requirements of semantic representation and simplicity to narrow the field of candidates. But we still have no *a priori* guarantee that only Chomskian grammars, with their particular constitution of phrase structure rules and transformational rules, will do the trick.

Still assuming, for the sake of argument, that Chomskian grammars are uniquely capable of meeting all these theoretical requirements, the triumph of this fact may be completely overshadowed by a certain logical necessity. Wittgenstein in his *Tractatus Logico-Philosophicus* and numerous other philosophers since have sought to establish by transcendental reflection on the nature of reality and thought that the only logical possibility for the structure of language is that of reference, identity, predication and quantification. Thus any language must somehow flesh out an austere skeleton of linguistic structure isomorphic to that of what logicians know as the language of first order logic. This is too deep a subject to go into here. We shall confine ourselves to the remark that Wittgenstein's philosophical argument could attract biological backing by considering the obvious truism that organisms with a sensory-motor system would do best to organise their worlds as though composed of objects (like trees and rocks), agents (like themselves and predators) and events (like pursuits and copulations). Reference to such items will in all probability be a primary function of any language they come to use. It may turn out that the linguistic structures of attribution, adverbial modification, verb tense and so on, all the way up through the upper reaches of semantic sophistication, are determined, logically, and biologically, as the only possible way of making sense of, and imposing order upon the world.

If this argument for categoricity of linguistic structure holds, then of course the hierarchical structure revealed by Chomskian grammar lies there waiting to be discovered. But the Chomskian thus takes the wraps off a logical necessity, and one will be inclined to ask what momentous *empirical* discovery has been made when it is claimed that human languages are cut to a special pattern – namely that, deep down, of the logicians' categorial languages.

This special pattern has empirical significance only if it can be mastered by mere mortals and not only by recourse to the most extensive conceivable class of computing devices. The 'mere' in 'mere mortals' requires some elaboration here. To survive, we must in some abstract sense be able to compute. But just how much computing power do we need to survive? Presumably evolution has endowed us with computing capacity 'from below'. Is it not unreasonable to suppose that our brains have to be modelled as the most powerful machines possible, and that nothing weaker will do?

This will no doubt be anathema to those theorists of mind like Lucas who have sought in Gödel's incompleteness theorem for formal systems of arithmetic vindication of the view that the human mind, since it is able to diagnose arithmetical truth for sentences that any computing device must pass over in whirring cybernetic silence, must thereby transcend the latter in computational power, insight, ingenuity or the like. (We discuss this claim in section 8 below.)

Our concern at present is not so much the question of what mental or intellectual power is needed in order to grasp arithmetical truth, but rather that of how much computational power is needed to diagnose strings as grammatical sentences, and to recover their meanings. We presume man coped with quite long sentences before he invented arithmetic. Let us assume that some lobe of his brain processes utterances for grammaticality, and let us call it the g-lobe. It need not be a particular chunk of grey matter, but could be constituted cybernetically by the multifarious participation of nerve cells and fibres within the whole brain. Our question now is: must the g-lobe be counted among the most powerful computing devices imaginable?

On evolutionary grounds, one would hope that the answer to this question would be negative. And there is presently a very interesting development in theoretical linguistics, in which Gerald Gazdar (1981) is a prominent figure, which promises to support this negative answer. Gazdar and his associates maintain that natural languages such as English are, in the terminology of mathematical linguistics, *context-free*. What this means is *not* that its sentences have a definite meaning independently of the context in which they may be used. It means rather that the phrase structure rules (of which the grammar consists exclusively, transformations being conspicuously unnecessary) are formulated in a way that allows a category label to be re-written without regard to the surrounding context of symbols in which it appears. In brief, the way branches develop within a deep-structural tree depends only on what is at the point of branching, and not on what lies else-where on other branches within the developing tree. By contrast the phrase structure rules of a *context-sensitive* language do allow for restrictions that require certain category-label configurations to be present before a particular category label may be re-written as a string of others.

Context-sensitive and context-free languages are *decidable* – that is, for languages L of these types, there are mechanical decision methods which one can apply to any given string to decide whether it is a sentence of L. Clearly this is what users of natural language do all the time. Speakers and hearers process sentences remarkably quickly, producing almost instantaneous verdicts on grammaticality. This would incline one to try to locate natural languages at least within the class of decidable languages. But Chomsky's theory of transformational grammar, as we have seen, does not do this.

There are some neat and interesting results in mathematical linguistics concerning the kind of computing device that can serve as a parser for context-sensitive languages, and context-free languages. (For a fuller treatment, the reader is referred to Hopcroft and Ullman, 1969.) We can imagine a parser for L as a black box into which we feed a tape with strings of lexical items to be tested for sentencehood. If the machine swallows a string on the tape and stops, then that string is a sentence of the language, and conversely; but if the machine 'chokes' and stops before swallowing the whole string, then that string is not a sentence of the language, and conversely. We now know that parsers for context-sensitive and context-free languages are of the following kind.

Context-sensitive

Linear bounded automata – essentially, machines that are allowed only a limited amount of 'scratch tape', bounded in length by a linear function of the length of the input string. Thus the shorter the string, so proportionally shorter the amount of memory tape needed in order to reach a verdict as to grammaticality.

Context-free

Pushdown automata. Such machines have a 'one-way' memory tape for scratch work, and can 'access' only the most recent entry on it; otherwise, accessing a distant entry entails loss of memory store since that entry. Crudely, such machines have very short-term memory; and can effectively recall past events (i.e. vividly remember) only when the recent past is of no consequence.

This so-called sub-recursive hierarchy of computing machines is *prima facie* more plausible company for the g-lobe than the heady heights of Turing machines with their potentially infinite tapes. For grammars should be parsers as well as generators. That is, they should produce decisions as to the grammaticality of presented strings as well as generate grammatical strings. Now Janssen, Kok and Meertens (1977) show that it is logically impossible both to enumerate a class of grammars each of which generates infinitely many sentences, and at the same time provide each infinite recursive (decidable) language with a grammar somewhere in the enumeration. Thus if we wish, with Chomsky, to insist on the potential infinity of sentences in any natural language, and at the same time heed common sense and the consensus of workers in the field of natural language processing by regarding natural languages as decidable (because parsable), we must view natural languages as a special sub-class of the decidable ones. This puts them down among the context-sensitive or the context-free languages.

As it turns out, Gazdar proposes that English is a context-free language. He ingeniously re-works the effect of extant transformational rules in Chomskian grammars for fragments of English by re-casting them as context-free rules. The implicit claim behind this project is that although Chomskians have so long insisted that they had found the right form of grammar for natural language, nevertheless they themselves have not availed themselves of the full power of those grammars, but have instead only been exploiting their lower reaches.

It is far too early yet to be able to pronounce on this interesting new trend in theoretical linguistics. But Gazdar's claim might fit more convincingly than Chomsky's into a picture of the gradual evolution of man's cognitive apparatus and cybernetic functions. Our brains do not have potentially infinite 'storage' (analogous to an automaton's scratch tape) when working out whether strings of sounds are grammatical. Parsing goes on in the head, and whatever is in the head is finite. That includes memory store. Thus automata that model the brain's linguistic abilities might be 'linear-bounded' or weaker.

The abstract account of computing devices provides for a memory tape and a black box scanner. We know from the work of McCulloch and Pitts (1943) that nerve nets can do duty for any such scanners. Now there is always the chance of a trade-off between size and complexity of the black box scanner (whether thought of in terms of the 'machine table' of state transitions, or in terms of the neural wiring of its physical realisation) and the available memory store for any given class of computing problems. For example, a class of problems soluble more quickly by a large and complex pushdown automaton – one 'low down' in the hierarchy – might be soluble more slowly by a small and simple linear bounded automaton that makes up for its size and simplicity by exploiting its tape allowance to the full when dealing with those problems. The relative simplicity and complexity referred to here, however, is that of the black box, *not* that of the physical realisation of the memory device.

Evolution has shaped our brains to compute fast and efficiently over short time spans, exploiting only a finite internal memory store. Only relatively recently has the use arisen of external memory aids, surely eschewed in parsing. It is such external aids, analogous to an automaton's scratch tape, that now invite one to compare our computing abilities with those of machines 'high up' in the hierarchy. But this comparison is misleading. Fed with infinite tapes, of course we could carry out the instructions of a Turing machine table and dutifully perform the computation it involves – provided we stayed alert and lived long enough. But interest in our computational capacities as language-users resides in the fact that the relevant computations – the mediations between input and output – take place entirely in the head, where no potentially infinite tape lies coiled. When a completed neuroscience has clarified the structure of axons, synapses, dendrites, ganglions and all the other delicate details of our grey matter involved both in memory and in the state transitions of the 'black box', we may be able to devise cybernetic models of overall performance better matching regular or pushdown or linear bounded automata than more powerful Turing machines. In this connection it will remain of crucial interest to try to determine how much computational power is involved in our ability to parse sentences of natural language.

We badly need a clear characterisation of the sorts of empirical test that could be used to decide the matter of a correct grammar in favour of Chomsky or any of his linguistic rivals. Indeed, such empirical tests are arguably impossible. Certainly, we can eliminate from contention any generative grammar that produces a patently ungrammatical sentence at any point. Furthermore, if the grammar not only generates, but also *decides* or *parses* given strings, it can be eliminated if it gives a patently false answer to the question whether a certain string is grammatical. But suppose we have either a generator or a parser that has thus far at least not made any outright mistakes. Are we to recognise it as *the* correct grammar? Certainly not. For we have no guarantee that it will not produce unacceptable results some time hence. All we know is that it has so far produced only correct outputs. We have only inductive evidence therefore, which is far from conclusive, that it will continue always to do so in the future.

Thus it is far from clear whether it makes sense to speak of *the* correct grammar waiting to be discovered by attentive observation and theoretical diagnosis on the part of the linguist; the same grammar, indeed, as he then wishes to regard the competent speaker as somehow having internalised or implicitly grasped.

Consider the possibility that many different grammars of radically different kinds produce as their output exactly the same class of, say, English sentences (or decisions about grammaticality in English of presented strings). How might one decide which is *the* correct one? One could perhaps appeal to further considerations over and above mere extensional agreement in output. One might require for example, that the grammar also assign semantic representations and account for the logical relations between sentences. One might further require that it be computationally simple or economic in generating its output, and also be of the kind that can be 'constructed' by a learner in a simple and economical fashion from the data at his disposal. But still there would be no *a priori* guarantee that Chomskian grammars are the only contenders left in the field.

On the account of the origins of hierarchical linguistic structure advanced by Sampson, especially with the substitution of Montague grammar for Chomskian, there is no need to appeal to innate linguistic abilities any more specialised than those required for general problem-solving. Subconsciously scientific, the child implicitly tests hypotheses about grammaticality against the linguistic data it encounters. Moreover, presumably because the child's programme of exposure (not to gross input, but to selectively filtered representations thereof) and learning mirrors that of the phylogeny of language itself, the hypothesis of hierarchicality will entrench itself very early on – for the language is indeed hierarchical! Thus by a Popperian process of hypothesis formation and testing, the child arrives at an adequate set of rules.

The Chomskian might advance the speculative thesis that any mutation causing children to search immediately in the right class of grammars would have a great selective advantage, and that such evolutionary change might well have taken place, producing human beings who are now pre-programmed to process linguistic data in a specific way. Certainly, the selective pressure for advanced linguistic competence would be very great – as witness the probable absence of any human language, in the intermediate range of the evolution that Sampson describes, that lacks, say, some of the syntactic and semantic functions and resources of known languages. But whether genes for hierarchical supposition would confer a sharp selective edge on language learners carrying them depends very much on how satisfactory an account of language acquisition is to be had from a suitably sophisticated (non-behaviourist) learning theory. Chomsky places faith in special neural mechanisms blossoming in the brain. He is ungroundedly pessimistic over the prospect of developing a powerful enough learning theory to account for language acquisition. His argument, if it counts as such, is a dogmatic admission of defeat, unsupported by quantitative evidence. We simply do not yet know enough about the capacity of heuristic based learning strategies that have been discovered or perhaps remain to be discovered (see Pinker, 1979).

8 THE MECHANISM OF MIND

We mentioned above that some philosophers – most notably J. R. Lucas (1964) – have tried to argue that the proof of Gödel's incompleteness theorem for arithmetic shows that the human mind (or brain, depending on one's views about the mind-body problem) transcends any Turing machine in its cognitive capacity. It is worth explaining further the line of argument they have employed here.

Gödel originally proved his result for the system of *Principia Mathematica* set forth by Whitehead and Russell in the year 1910. This is a system of axioms and rules of inference allowing the formation of proofs of arithmetical statements. No falsehoods of arithmetic can be proved in the system, and Russell and Whitehead had simply assumed that it would also be the case that every *truth* of arithmetic could be proved in the system. Gödel's famous result showed them to be wrong. He first of all showed that the linguistic expressions employed in the system could be coded as whole numbers. Then he showed that all the important relations among expressions that are involved in checking that a given construction is a proof in the system determined relations among their code numbers of a very special kind, called primitive recursive relations.

The next step was to show that every primitive recursive relation is representable in the system, in the following sense. Suppose R(x, y) is a primitive recursive relation between pairs of numbers x, y. Then there is a (complex) formula $\psi(x, y)$ in the language of arithmetic such that, roughly, R holds just between those numbers x, y for which the statement $\psi(x, y)$ is provable in the system. Thus the primitive recursive relational facts about expressions of the language are represented, via the coding, by provable arithmetical relations between the corresponding code numbers in the system. (For a fuller treatment, the reader is referred to Tennant (1978).) In this fashion the system can, in a roundabout fashion, 'talk about its own syntax'. Omitting now the complicated details of the last step of Gödel's proof, we shall simply summarise it by saying that he was able to proceed to the construction of a sentence that, in effect, asserted its own unprovability. The sentence G involved had the code number γ, and was of the form 'for all numbers x, x is not the number of a proof of the sentence with code number γ'. Hence every instance of the universal claim G is true, but G itself is unprovable. But surely, if every instance is true, so must be the universal generalisation? In other words we, as users of the formal system, placed 'outside' it, can see that G must be true, while the system itself is, as it were, blind to the fact. Our arithmetical capacities importantly exceed those of *Principia Mathematica*.

What now of the claim that we transcend any given Turing machine? Lucas's argument is as follows. Let any machine M be given. M, in order to be taken as a serious rival to our own mental capacities, must have built into it an 'arithmetical system'. That is, on its print-out tape there must be a suitably coded list of the code numbers of sentences of arithmetic that it regards as theorems. Now, *if* we can discern from the workings of the machine just what the axioms and rules of inference are by which the machine generates its own repertoire of arithmetical truths, then we can apply Gödel's construction to obtain a sentence G_M that we can see to be true, but which is unprovable in the machine's own system of arithmetic. But this is a big 'if'. Certainly, any machine that ingenuously discloses its generative system for arithmetic can be 'out-Gödelised' by us, the insightful constructors of undecidable sentences. But it is not true in general that we can detect the axioms and rules and so come up with the required sentence. Thus Lucas's argument has an important lacuna, and we do not regard it as valid.

Gödel's result has been refined and generalised in the last few decades of foundational research. It can now be stated without reference to any system of axioms and rules such as *Principia Mathematica*. In its purest form, it states that any true arithmetical theory, all of whose sentences can be listed by a Turing machine, will lack at least one truth of arithmetic. More briefly, the set of arithmetical truths is not axiomatisable. In this sense of 'axiomatisable' there are all sorts of mechanical recipes for listing sentences that do *not* proceed from axioms to reach their theorems by following rules of inference. Gödel's argument, however, does not provide for such a 'non-axiomatic' system any definite sentence that we can put our fingers on as the one missing from the list, thereby proving our superiority. Only certain very well-behaved and orderly systems, namely those set up in terms of axioms and rules of inference, enable the construction of definite absentee truths. It is no use, therefore, to claim that given any Turing machine there is a sentence that we can see to be true but which the machine cannot. For in the case of 'non-axiomatic' machines, the mechanist can ask 'Exactly *which* sentence is it that you have in mind?' And faced with this question the Lucasian must confess his ignorance. We can even drive the point home by saying that human arithmeticians are no more than Turing machines who, despite the famous attempts of great representatives like Peano, Dedekind, Russell and others to write down definitive descriptions of the sources of their perception of arithmetical truth, have failed to describe exactly what 'makes them tick'. Since Gödel out-Gödelised *Principia Mathematica*, certainly Gödel's arithmetical sub-component (considering him as an automaton) is different from (perhaps definitely stronger than) *Principia Mathematica*. We are, however, far from knowing what that arithmetical sub-component is. Nevertheless it is still permissible to regard it as no more powerful than a Turing machine. Being unspecified, it is presumably immune to Gödelisation. Of course, it is also incomplete, just as *Principia Mathematica* is. But even if some agent X *could* correctly work out what that sub-component is, and definitively characterise it as a finite generative system, and out-Gödelise it, this remarkable state of affairs would still go no way to showing that X transcended all Turing machines. X could be a Turing machine or something even weaker, like the rest of us. Moreover, his success at out-Gödelising even the arithmetical sub-component of Gödel would of necessity be a localised one. For no Turing machine is able to perform the general task of specifying, for any given machine, what its arithmetical component is. Thus no Turing machine can out-Gödelise all others; perhaps the best argument for egalitarianism there is!

Speaking of human brains as Turing machines contains an important grain of truth, but in some respects it is strained and far-fetched. For the formal definition of a Turing machine is of a finite state machine programmed to perform operations from a finite repertoire as it processes inputs and produces outputs from a finite alphabet on an *infinite* tape. With all the scratch paper in the world, and a whole life devoted to it, there are obviously tasks that some Turing machines can perform in principle which we would never perform in practice. But our predicament is only that of mortality and limited resources. The central thesis of artificial intelligence is that all our cognitive processes are in principle simulable by Turing machines. We just happen to go on the blink, wear out and die.

Could this very finitude, however, allow us to push human cognitive capacities further down in the hierarchy of mechanism? We have already adverted to this possibility above, in connection with the generative powers of adequate grammars. If this could be done across the whole range of our capacities we might be edging towards a cybernetic materialism within easier evolutionary reach. To settle this question will involve, among other things, looking at the particular tasks that have been shown to be soluble or calculable by machines in one class but not by any machine lower down in another class. It may turn out that all these 'impossibility' proofs present fatal problems for the machines lower down *only because of their sheer length or size*; and that, moreover, they are soluble by the stronger machines only because the latter are allowed unlimited time and resources to compute the required results or answers within the constraints of their own types of programmes. For example, the set of all strings of zero's of prime length is not a context free language, and so cannot be parsed (accepted, recognised) by a pushdown automaton. But, given any n, the (finitely) many strings of zero's of length less than n *does* form a context-free language.

In the finitary predicament, could this phenomenon turn out to be general? We have finite brains, finite resources, finite storage and finite computing time. What we can do and what we actually do should therefore be representable – like the simple example of the prime language with an upper bound on sentence length – by 'weak' finite automata.

We cannot see any reason other than theoretical convenience for the mechanist, once he has won his long-standing debate with the mentalist, to have to resort to Turing machines in general. Why all the emphasis on whether the mind is a *Turing* machine? If we decide it is, shouldn't the realistic next step be to demote it a rung or two?

Perhaps the only reason why cybernetics dwells in the upper reaches of the recursive hierarchy when modelling cognitive capacities is that it issues in simpler models, terser programmes and more manageable modules. The real story, however, in all its flickering and evanescent complexity, might be that of an immensely large but sub-recursive automaton.

We say 'immensely large', meaning not that our heads are swollen but rather that the brain is capable of being in an immensely large number of states. There are 10^{10} neurons in each human brain. The number of interconnections and firing configurations is staggering. No one could even begin to gain insightful understanding of correspondingly large finite automata in the abstract in trying to map our own cognitive capacities. So we abstract further, to the more powerful programme loops of the strongest machines, to bring manageable order to the confusion.

9 INTERPRETATION OF APE LANGUAGE EXPERIMENTS

Several experimenters have maintained that chimpanzees and gorillas can be taught to communicate by means of artificial systems not involving speech. For anatomical reasons, chimps and gorillas cannot produce a range of sounds sufficient for speech. But some investigators claim that they have taught them to use systems approaching human language in their versatility and complexity. The implicit claim to be considered is that apes are, qualitatively speaking, a quantum jump ahead of other animals in this respect (although admittedly not many others have been investigated): that they are 'on our side of the divide'.

The names of those individual apes singled out for attempted communion are now legion. The Hayes tried first to introduce the chimpanzee Viki to the spoken word. Cathy Hayes (1952) gives a very non-anthropomorphic assessment of how much Viki understood, and a reasonable explanation of her successful responses to verbal prompting through routine action sequences. A (rather oversimplified) version of American Sign Language was the method of communication attempted by Patterson with the gorilla Koko, and by the Gardeners with the chimpanzee Washoe. Terrace and his associates chose to try Pidgin Sign English on their chimpanzee Nim Chimpsky. Premack's exchanges with the chimpanzee Sara involved plastic symbols; while the Rumbaugh's programme with the chimpanzee Lana involved a computer lexigram language known as Yerkish. A full account of their projects can be pursued through our bibliographical references. The claims initially made by ape language experimenters on behalf of their subjects have recently been critically re-assessed (see especially Seidenberg and Petitto, 1979).

The criticism is of various kinds. Some concerns the basic methodological pitfalls surrounding any project of this kind – the lack of proper controls, and over-readiness to read richer interpretations into bits of behaviour than a more rigorous viewpoint would endorse. Another set of criticisms concerns the basic approach, given the desired object of establishing communicative contact with and among the apes concerned. Anyone can see that artificial keyboard or lexigram languages can be imparted, if at all, in only the most impoverished social situations – quite unlike the nexus of warm and intimate physical bonds that probably prevailed at the dawn of language millennia ago. Normal communication is free and spontaneous, with parties taking turns, with an equal balance between initiative and response. A human child quickly expands his vocabulary, and the mean length of his utterances increases accordingly. He uses already mastered words in new combinations to which he has not hitherto been exposed, and responds to such combinations appropriately as well.

Other criticisms focus on massaging of the data, in the form of simplification of ape utterances in their reporting, so as (probably unwittingly) to make them look more like human utterances than the hodge-podges they really are; and re-analysis of unedited films of 'discussions' with the apes, showing just what a high proportion of the exchanges embody no more than unconscious cueing by the human experimenter, slavish imitation by the ape, or his redundant expansion and embroidery of their exchanges without commensurate amplification of information conveyed. More philosophical criticism is to be found along Davidsonian lines. Apes may well have produced new two-sign combinations that their trainers have been inclined to interpret as appropriately invented for some feature in the context. A celebrated example is Washoe's signing of 'water bird' in the presence of a swan. But the radical interpreter could re-interpret the dyadic sign as 'bird on water' or 'wet feathers' or 'reflection of bird' or in any one of many other ways. The dyadic sign need not be endowed with any unitary significance, or be regarded as a newly constructed compound of previously mastered components. On this view it would be overcharitable to credit ape strings with syntactic structure when they may be nothing but sequences of single 'word-sentences'.

As we noted in our discussion of generality, tense, etc., it is easy to underestimate just how difficult it is for attributions of attitude and meaning to pass methodological muster. Davidson enjoins the radical interpreter to be *nasty* in thinking up as many competing interpretations of observed behaviour as possible. A good example of how nasty one can be is to be had from the chimpanzee language experiments. A scholarly audience will forgive this quote from *The Listener*:

> The Gardeners first taught Washoe by shaping her hands for each sign. Hands together means – 'more'. When Washoe got close, she was rewarded by getting more of what she wanted. Soon she was using the sign for more of anything – food, rides, and so on – proof that she had grasped the general concept of the word. (Leakey, 1981b)

Let us be nasty towards Washoe. The word 'more' has its meaning disclosed through combination with terms telling one what it is of which 'more' is in question. No doubt the word as used by infants will occur on its own, without the added term, and adults satisfying the need will work out the missing term from context. But adults respond linguistically too, supplying the missing substantive with the wanted substance. The child in due course in new contexts uses the modifier with substantives, thereby expressing his desires more clearly. The process of learning words, learning their grammatical categories and acquiring them in correct combinations is very much a two-way affair. It involves requests, solicitous enquiries, offerings, forbiddings and factual commentary. But does Washoe use the hand signal, allegedly meaning 'more' in this way? Does she offer more of things to her interlocutors when 'asked', as well as demand things from them? Does she perform several different speech acts with the word, questioning, commanding, wishing, stating? Watching the film of Washoe with Leakey's commentary, the nasty radical interpreter is struck by any number of competing interpretations of Washoe's signing. Why not interpret her as expressing something tantamount to 'I have an unfulfilled desire', or, 'make me feel nice!'? Either of these – quite unconnected with 'more' of anything – would cohere with the facts just as well as the 'more' hypothesis and explain the same range of behaviour. If Washoe has an unfulfilled desire, one can rest assured that the eager and attentive experimenters will gather from context what it is and fulfil it if at all possible. Nor is it unlikely that the formation of the desire will be closely linked with current experiences and will therefore probably be fulfilled by providing more of whatever is salient in the context – porridge, or rides in a plastic tub, etc. Likewise the hand sign might mean 'continue' or 'would that things remain roughly as they are now, in lumps or in intervals'.

10 HUMAN LANGUAGE ACQUISITION

One of the most important points of difference between language acquisition by human beings and the deficient parallel process in the case of the ape language experiments is, of course, that in the latter case one is not dealing with the acquisition by the young of a system employed already by conspecific elders. The apes being taught are therefore without an evolutionarily conferred advantage that human children enjoy – that of employing learning techniques, and being initiated by their elders, in a way that has presumably been refined by selection pressures over a very long time. Recent studies of language acquisition reveal that children are very attentive, and actively process evidence in ways perhaps not fully appreciated at the time when Chomsky was championing a theory of innate linguistic universals that depend heavily on an alleged gap between the scanty data available to the child and the rich system that he eventually masters in response thereto. Children are highly motivated learners of language, a disposition probably by now 'wired in' to our species. In this respect they are crucially quite unlike the apes, who have not yet been subjected to evolutionary pressures for rapid acquisition of symbol systems. Moreover, there is a contemporary school of research (Trevarthen, 1979) that is bringing home through painstaking and meticulous observation of infants with their mothers, just how rich is the rapport between the two. They have coined the term 'intersubjectivity' to cover the subtle and instinctual goings-on, from the earliest days of infancy, that are, it seems part of an innate programme guiding the bonding of mother to child, and her unconscious cueing of its cognitive and linguistic development. Children benefit also from more than usually grammatical speech from adults who address them in the early stages in a fashion tailored to their learning needs. This tendency on the part of adults could of course also be selected for, once language was entrenched enough to exert the required pressures. Parents repeat and reinforce their children's utterances, and produce slightly more complicated versions of things already mastered in a gentle advance up the ontogenetic slope (see Marshall, 1980).

These observations on human language learning, coupled with evolutionary speculations, make one realise just what a formidable accomplishment it would be if apes could be taught the use of a communication system remotely approximating a natural language in creativity, recursiveness, and the extreme conventionality revealed in such matters as reference to spatio-temporally remote items or counterfactual conditionalising or universal generalisation about an unsurveyed domain. They may well have certain cerebral pre-adaptations that subtended vocalised speech linked with gesture, which responded in some ancestral line to pressures of selection for an ever more complex code of communication. Indeed, Jaynes (1976) claims that this happened as recently as the late Pleistocene. The proliferation of tools in the Pleistocene, the making of apparently 'non-utilitarian' and symbolic objects, the practice of burying the dead ceremonially, the construction of shelters and the marked increase in the size of the frontal lobe of the brain with Broca's area, all point to language's being reasonably developed by the end of the Pleistocene (see the papers by Isaac, Marshack and Holloway in Harnad *et al.*, 1976). The ape's failure to acquire a 'language' from us (spoken or gestural or plastic-symbolic) in no way discredits the very reasonable evolutionary claim that they, or a common ancestor of ours, had the rudimentary cerebral beginnings that are now our speech centres. The language faculty is closely tied up, both conceptually and empirically, to other cognitive faculties such as silent foresight, planning, anticipation, reasoned fear, etc. It may be impossible for the apes to master a code that we would translate into our own terms. Admittedly we can discern glimmerings of intelligent thought involving the location of sub-goals and the execution of sub-strategies, a grasp of 'causality', and a keen grasp of the internal motivational states of others, as disclosed by their behaviour; and some experimenters (Suarez and Gallup, 1981) have even looked for signs of a sense of self. Perhaps soon they will discern ape intimations of mortality. But too much mist obscures the question what it is like to be a chimp for us to succeed in our best-meaning efforts to make them make the best of meaning: at least, in the full-blown human sense.

11 NEUROLOGICAL CONSIDERATIONS CONCERNING LANGUAGE

Our interpretation of the results of the chimp language experiments points to recursively structured language as a unique accomplishment of our own species. Lenneberg (1966) has argued that a great deal of evidence about deafness in children, aphasias, environmental deprivation of various kinds, muscular debilities and so on, supports the hypothesis that language acquisition by children follows a definite maturational path, passing various milestones of achievement in a certain order. Moreover, these milestones correspond broadly with others in the course of sensory-motor development, and are found in step with them even when the whole process is slowed down, as in the cases of mental retardation. Both linguistic and sensory motor development are co-ordinated further with the 'brain maturation curve'. This charts degree of organisation and certain chemical concentrations. There is no evidence of cerebral 'rubicons' that correspond to the different stages of language acquisition, but one cannot avoid the impression that the linguistic skills acquired by the growing child are orchestrated by a physiological score; and one, moreover, that has its own peculiar, species-specific crescendoes.

Apes and monkeys employ a limited number of calls. Some may serve to warn fellows about very specific kinds of predator, and therefore have quite specialised communicative significance (Seyfarth *et al.*, 1980). But they are not formed into sequences having different significance according to their manner of combination and the basic items they contain.

There are some noteworthy differences in brain physiology, apart from the massive increase in brain size as one passes up through the other primates to man. Chimpanzees lack Broca's area (for muscular co-ordination in speech), as well as the neighbouring arcuate fasciculus. The latter is the fibre bundle connecting the auditory area with Broca's area, which some believe to be crucially involved in the imitation of sounds. It is little wonder then, given these handicaps of sheer anatomy, that chimps are incapable of speech. A more delicate question to assess, however, is whether one could reasonably maintain that they are preadapted for *Language* (not necessarily vocal, but symbolic and semantic in an appropriate sense). We appear to share with chimpanzees a strong rooting of emotional cries in the limbic system – the so-called 'emotional' brain. But for the rest of spoken language, our neo-cortex is heavily involved, with strong evidence, furthermore, of lateralisation for speech even as early as birth. (By 'lateralisation' here we mean that the functions are based on one side – usually the left – rather than the other.) Infants appear, on recent experimental evidence, to be 'wired' for the recognition of speech sounds as opposed to others. And yet, paradoxically, the evidence for lateralisation of the 'higher' and more abstract processes of language apart from vocalisation, is less conclusive.

Geschwind (1974) once proposed a theory concerning association areas in the brain. According to him, language could only evolve once the auditory and visual *association areas*, that in monkeys as well as man convey impulses from the auditory and visual areas to the limbic system, *themselves* acquired suitable connections via yet another association area, thereby making possible so-called non-limbic cross-model associations. This 'association area for the association areas' is the angular gyrus, and it is absent in monkeys. The reason why its mediation of modes is so crucial is that learning and understanding names of objects probably requires the association of visual with sound images (assuming, of course, that we are dealing with names from a *spoken* language). Of course, naming is not the whole of language, as the opening passage of Wittgenstein's *Investigations* makes us so well aware; yet it is no doubt a central part of language, and still an important and controversial topic in the philosophy of language. Geschwind's express intention was to find physiological correlates of linguistic capacities, in order to bring language within the scope of scientific materialism. His conjecture, then, is just one way of making sense of varying anatomical facts across species, evolutionary hypotheses, and observations of impairments of linguistic functions on the part of human patients who had suffered different sorts of damage to their brains.

Surprisingly, Dingwall (1975) states categorically that Geschwind's theory, 'while ingenious, is most assuredly incorrect'. He cites experiments that have shown that

> Chimpanzees are able to match visually presented to tactilely presented stimuli . . . chimpanzees can transfer from auditory to Ameslan signs in the absence of referents. . . . Thus, the inability to produce cross-modal associations is not a barrier to the acquisition of language in chimpanzees . . . Not only is there evidence of similarity in cross-modal association behaviour in chimpanzees and man but also, as pointed out . . . there is no marked difference in these two species in the angular gyrus which

> Geschwind holds to be important in mediating this behaviour. (p. 43)

This is extremely puzzling argumentation. Chimpanzees have angular gyri. But given the copious evidence that chimps simply can't get their lips and tongues around enough sounds, their cross-modal associations pose no threat to Geschwind's theory. On the contrary, the very successes claimed by the chimp language experimenters, to which Dingwall himself is a sympathetic, would bear out that theory even further. For Geschwind had been careful enough to note that the angular gyrus is needed even for visual-visual (indeed, for any non-limbic) associations. So perhaps it is because they have angular gyri that chimpanzees have been able to sign successfully. We say 'perhaps' because rhesus monkeys have recently been claimed to be able to make cross-modal association (Cowey and Weiskrantz, 1975). Nevertheless, so far only chimpanzees have displayed any ability to make visual-auditory associations (Fouts *et al.*, 1976). And it remains to enquire to what extent other associative abilities experimentally revealed in other species approach those to be expected of a creature able to use linguistic symbols. Von Glaserfeld (1976) has argued that:

> the semanticity of a *linguistic* sign is constituted not by a tie that links it to a 'thing', but by one that links it to a representation or concept. The fact that a sign, be it verbal or non-verbal, has acquired symbolicity, does of course not preclude that it still be used as a perception-bound sign whenever there is a perceptual input that corresponds to the representation it designates; nor does it preclude that it be used by the sender to trigger a conventional active response in the receiver (as in the case of an 'imperative'). But what gives a sign the status of symbol is that it *can* be used without such a stimulus and without triggering the active response. (p. 222; his emphases)

In the same Lockean spirit, Davenport (1976) reaffirms the relevance of non-limbic associations when he asks:

> Of what relevance is cross-modal perception to the origin and evolution of speech and language? First, it appears that multi-modal information extraction of environmental information is likely to result in more veridical perception, and may facilitate cognitive functioning. Second, in my view, cross-modal perception requires

> the derivation of modality-free information, a 'representation'. That an organism can have the same representations, concepts or precepts, regardless of the method of peripheral reception, confers a great advantage on that animal in coping with the demands of living. . . .
> To the extent that an organism has a 'tag' for the representation, be it sound, gesture or combination, the process would be greatly facilitated. (pp. 147-8)

Just as there appear to be no rubicons in a child's cerebral development, so also there were probably no such rubicons phylogenetically. Evolutionary pressures would have gone to work on whatever genetic variability there was in the ability to make the relevant cross-modal associations underlying linguistic competence. The importance of the former for the latter is not at all diminished by exhibiting languageless creatures that can associate across modes. For, that one must be able to do X in order to do Y does not entail that if one can do X then one should be able to do Y.

Are the recursive resources of language an innate endowment of *Homo sapiens* alone? Our earlier discussion inclines one to say so. But we must not support the innateness claim with the wrong arguments. Universality is no guarantee of innateness (compare a similar point about ethical theorising in chapter II). This is so even though it might be true that any infant, regardless of race or family niche, can acquire the tongue of any community. This acquisition claim has never been rigorously tested. Physiognomical differences between racial types might affect pronunciation of the foster mother tongue. But setting purity of pronunciation aside, could any child master any first language so well that native speakers would regard him as fully competent in matters lexical and grammatical? Should languages have diverged sufficiently radically in our evolutionary past, and should cerebral organisation have adapted constantly to the needs of parsing, it is quite possible that counter-examples to the claim of eventual competence might be found. But this would not count in any way against language's being innate. Indeed, it is its very innateness, given sufficient divergence between different languages, that make such counter-examples seem possible.

12 GLOTTOGONIC THEORIES

Hewes (1977) counts twenty-four different theories concerning the origin of the first lexical items. (Perhaps 'theory' is too grand a label for these suggestions.) He concludes plausibly that the best account will probably incorporate several different suggestions about the origin of words in different 'lexical domains'. Onomatopoeia is one such suggestion.

The composer Leonard Bernstein has also suggested in all seriousness before a Harvard audience that introspection had delivered the remarkable association of 'gr'-sounds with the concept 'large' – as in 'gross' (German) and 'grand' (French): This theory is known as the ding-dong theory – to the effect that man resonates on an appropriate note after contact with a concept – and is due originally to F. Max Müller. As mysterious as it may seem, there is recent evidence that Americans, when confronted both with an English word they know and a list of exotic translates in foreign languages, can select the right translate with a probability significantly above chance.

Another suggestion is the babble-luck theory. It has it that words were adopted after versions of them had been produced by babbling babies in highly salient circumstances, which led to the adults' adopting them as standard vocalisations. Most readers will no doubt also know of cases where parents will preserve an arcane toddler word in the family idiolect even when its meaning is quite opaque even to neighbours and friends.

But by far the most likely theory looks to gesture as the raw public material from which human communication grew. Speech could have grafted on to gesture once the vocal tract had developed sufficiently, with selective advantages that Darwin had remarked on. The hands are free when one talks; and one can be heard in the dark and in all directions, under cover and so on. Consistent with this theory is the synchronisation reported by MacNeill between hand movements and speech segments. He observes also that gestures can extend and replace full imitation, while yet being sufficiently iconic to secure uptake. Finally, although Neanderthal man had once been thought incapable of producing a reasonable range of vocal sounds, the fossil reconstruction on which this claim was based has been severely criticised recently (Du Brul, 1979).

In the case of natural language, natives' pre-formal agreement on the boundaries of grammatical division is the most important kind of evidence the linguist can glean as he searches for the grammatical recipes and ingredients of sentence meaning. One attraction of the gestural theory of language origin is that significant structure does not have to await words. Susan Goldin-Meadow's subjects were unacquainted deaf children; but they had normal parents who did not try to communicate with them by gesture, or at least not in *sequences* as the children did:

> the deaf children were seen to have devised their own manual signs. Many of them were based on actual movements. And they combined signs into sequences with relatively stable 'word' order preference. The most rapidly developing child eventually made use of such sophisticated devices as clausal embeddings. The steps in the development of these sign systems occur in a regular sequence, and at times which resemble the development of vocal speech by hearing children. (MacNeill, 1979, p. 720)

Of course, these children might have been interpreted according to overly lax canons – compare our discussion of chimp gesture above. But *prima facie* the generous semantic account of their project is much more plausible, given that they are agents just like us except in so far as they cannot hear. We should not regard their 'invention' of a structured gestural language as grounds for believing that earlier hominid hand-waving could thus rapidly have attained parsable results. After all, these children are presumably 'wired' for language. They will possess the same cognitive structures and whatever maturational schedules thereof that underlie hearing children's language acquisition. As we remarked earlier, their 'problem' was to find a channel of communication. The problem facing our forebears was to discover communication.

Gestural messages, so it seems, can have significant structure. Thus the structure of spoken sentences need not have derived entirely from combination of words that had replaced single gesture types. Instead, structured gestural sequences, or *syntagmata* as MacNeill (1979) calls them, might already have had a rudimentary grammar before they were overlaid by speech.

Whatever the relative order of gesture, speech and structure thereof, our quasi-conceptual evolutionary thesis would still be pressing: iterable contribution to significant structure across the whole language, be it gestural or symbolic, is almost certain to have needed a shrewd intentional grasp by communicators of what they were up to. Note that this claim would not be undermined by showing that infants 'mindlessly' master the communicative content of structured sentences before it dawns on them gradually that assertions express beliefs and that deceit is possible. For selective pressures for linguistic ability could easily reverse in ontogeny the order we maintain would be needed in phylogeny. Selection would have operated within the environment of an already structured code. Mastery of that code by distant descendants could therefore become more 'mindless' in the early stages (even those involving significant structure) as brains were shaped by natural selection for more and more rapid language acquisition.

13 SUMMARY

Our theory has, briefly, emerged as this. Language has obviously evolved. Noises and gestures among social creatures were probably the raw material. These creatures could have behaved in such ways as to merit attributions (by a human observer, could one have been present then!) of mental states such as perceptions, beliefs, intentions and desires, *before* they evolved a language. There is today a well-worked out philosophical analysis of how languageless creatures could attain to their first acts of a communicative nature. Here 'communicative' means more than just signalling, something more like Grice's non-natural occasion meanings. The propositional contents of these acts would have to be very basic. Conventions could have evolved in the way described by Lewis, without convening and without a language to decide upon them. These conventions would confer upon various noise- (cum gesture-) types certain relatively fixed meanings. Chimp language, natural or taught, is vaguely in this court. But somehow natural language became powerfully different. Somehow syntactic structure evolved, and with it a system of far greater generative power and so-called 'creativity' than before. In this connection the model of linguistic evolution due to Simon and Sampson deserves serious consideration. We have suggested in an Haeckelian spirit that one might find in Montague grammar the synchronic re-capitulation in syntax of the evolved complexities predicted by the Simon-Sampson theory. At this point many methodological and mathematical questions in theoretical linguistics have to be settled as well as questions about the interpretation of evidence. What is to count as evidence for a theoretical grammar? Why must language be 'recursive' from a theoretical point of view? Is grammaticality a decidable notion? Is the competence-performance distinction so easily made? Should grammars generate or parse? Is the ontology of the theoretical linguist to be that of the native speaker? What are linguistic intuitions? Might they belie an inchoate and only accidentally true theory of linguistic structure? What are the distinctive 'design features' of human language that sets it apart – if at all – from other animal communication systems? What is the right interpretation, in this respect, of the recent progress allegedly made by apes in coining new 'sentences' in the languages they have been taught? Are data about brain structure, compartmentalisation, layered evolution, etc., relevant here?

There are more questions here than we have discussed, and even more than we have tried to answer. The general upshot, we hope, is that we have shown how language could have evolved and contain its own fossil record, so to speak, in its grammar, in which the fulgurations are concealed and philosophical analysis is needed to locate them at significant loci in the phylogeny of language. The results of such analysis would have to mesh with data and theories about the maturation of competence and order of acquisition in the child. An important question is whether learning a language falls in the domain of general problem-solving – following a method of (non-conscious) conjecture and refutation, constrained only by the very broadest of search heuristics – or comes about because of a highly specific, innate or inborn cognitive structure that follows very specific strategies of acquisition in a characteristic maturational programme.

With phylogeny misty and ontogeny controversial, one can expect the present plateau of linguistic attainment – the course of which leaves no paleontological record – to produce a closed circle of radical interpretation. Meanings are parasitic, from the evolutionary point of view, upon movement and noise. But latterly meanings have become so spliced with movement, and of course causally inter-related, that it is all too easy to see the last twist of the human spiral, that took man through the birth of co-operative culture, in food-sharing, tool-making, building and hunting, as a closed circle. As learners of language, or as translators and interpreters, we now find ourselves in an overly synchronic predicament. The web of belief and fronds of desire are all too filigree now that language is available for their further refinement and extension into the purely linguistic domains of intended meaning.

BIBLIOGRAPHY

Adorno, T. W., *et al.* (1950), *The Authoritarian Personality*, New York: Harper & Row.

Ajdukiewicz, W. (1935), 'Die syntaktische Konnenität', *Studia Philosophica*, 1, pp. 1-27.

Alberts, B. *et al.* (1983), *Molecular Biology of the Cell*, New York: Garland.

Alexander, R. D. (1979), *Darwinism and Human Affairs*, London: Pitman Publishing.

Alexander, R. D., *et al.* (1979), 'Sexual Dimorphisms and Breeding Systems in Pinnipeds, Ungulates, Primates and Humans', in Chagnon and Irons (eds) (1979).

Alexander, R. D. and Noonan, K. M. (1979), 'Concealment of Ovulation, Parental Care and Human Social Evolution', in Chagnon and Irons (eds) (1979).

Alexander, R. D. and Tinkle, D. W. (eds) (1981), *Natural Selection and Social Behaviour*, New York: Chiron Press.

Anscombe, G. 'The First Person', in Guttenplan (ed.), (1965).

Armstrong, D. M. (1971), 'Meaning and Communication', *Philosophical Review*, 80, pp. 427-47.

Axelrod, R. and Hamilton, W. D. (1981), 'The Evolution of Co-operation', *Science*, 211, pp. 1390-6.

Ayala, F. J. (1974), 'Biological Evolution: Natural Selection or Random Walk?', *American Science*, 62, p. 692.

Baker, J. R. (1974), *Race*, London: Oxford University Press.

Barlow, G. (1980), 'The Development of Sociobiology: a Biologist's Perspective', in G. Barlow and J. Silverberg (eds), *Sociobiology: Beyond Nature/Nurture*? Boulder, Colorado: Westview.

Barnett, S. A. (ed.) (1958). *A Century of Darwin*, London: Heinemann.

Benhar, E. E., Carlton, P. L. and Samuel, D. (1975), 'A Search for Mirror-Image Reinforcement and Self-Recognition in the Baboons (Papio anubis)', in S. Kondo, M. Kawai and E. Ehara (eds),

Contemporary Primatology, Basel: Karger.
Bennett, J. (1976), *Linguistic Behaviour*, Cambridge University Press.
Bernstein, B., *et al.* (1971-5), *Class, Codes and Control* (vol. 1, 1971; vol. 2, 1973; vol. 3, 1975), London: Routledge & Kegan Paul.
Bertalanffy, L. von (1972), 'The Model of Open Systems: Beyond Molecular Biology', in A. D. Breck and W. Yourgrau (eds), *Biology, History and Natural Philosophy*, New York: Plenum.
Bitterman, E. R. (1965), 'The Evolution of Intelligence', *Scientific American*, 212, pp. 92-100.
Blaustein, A. R. and O'Hara, R. K. (1981), 'Genetic Control for Sibling Recognition?', *Nature*, 209, pp. 246-8.
Blick, J. (1977), 'Selection for Traits Which Lower Individual Reproduction', *Journal of Theoretical Biology*, 67, pp. 597-601.
Blum, M. S. and Blum, N. A. (1979), *Sexual Selection and Reproductive Competition in Insects*, New York: Academic Press.
Boswell, J. (1953), *Life of Johnson*, Oxford University Press.
Bouchard, T. J. and McGue, M. (1981), 'Familial Studies of Intelligence: A Review', *Science*, 212, pp. 1055-9.
Bower, T. G. R. (1976), 'Repetitive Processes in Child Development', *Science*, 235, pp. 38-47.
Braggio, J. T. *et. al.* (1982), 'Logical and Illogical Errors Made by Apes and Children on a Cognitive Task', *Journal of Human Evolution*, 11, pp. 159-69.
Brent, L., *et al.* (1981), 'Supposed Lamarckian Inheritance of Immunological Tolerance', *Nature*, 290, pp. 508-12.
Broda, E. (1977), 'Free Will and Evolution' (Comments on Young's Article), *Journal of Theoretical Biology*, 69, pp. 761-6.
Bronowski, J. (1970), 'New Concepts in the Evolution of Complexity: Stratified Stability and Unbounded Plans', *Zygon*, 5, pp. 18-35.
Burhoe, R. W. (1967), 'Five Steps in the Evolution of Man's Knowledge of Good and Evil', *Zygon*, 2, pp. 77-96.
Bush, G. L. (1975), 'Modes of Animal Speciation', *Ann. Rev. Ecol. System*, 6, pp. 339-64.
Campbell, B. (1974), 'The Physical Basis of Language Use in Primates', in M. L. Lead (ed.), *Frontiers of Anthropology*, New York: Van Nostrand.
Campbell, C. (1972), *Sexual Selection and the Descent of Man 1871-1971*, Chicago: Aldine.
Campbell, D. T. (1966), 'Evolutionary Epistemology', in P. A. Schilpp (ed.), *The Philosophy of Karl Popper*, La Salle: Open Court Publishing.
Campbell, D. T. (1975), 'On the conflicts between biological and social evolution and between psychology and moral tradition', *American Psychologist*, 30, pp. 1103-26.
Caplan, D. (ed.) (1980), *Biological Studies of Mental Processes*, Cambridge, Mass.: MIT Press.
Carlson, E. A. (1966), *The Gene: A Critical History*, Philadelphia: W. B. Saunders Group.
Cattell, R. B. (1972), *A New Morality from Science: Beyondism*, New

York: Pergamon Press.

Chagnon, N. A. and Irons, W. (eds) (1979), *Evolutionary Biology and Human Social Behaviour*, North Scituate: Duxbury Press.

Cherfas, J. (1981), 'The battle of the sexes: how to keep the numbers equal', *New Scientist*, 91, pp. 87-8.

Cherfas, J. and Gribbin, J. (1981), 'The Molecular Making of Mankind', *New Scientist*, 91, pp. 518-21.

Chevalier-Skolnikoff, S. (1976), 'The Ontogeny of Primate Intelligence and its Implications for Communicative Potential: a Preliminary Report', *Annals of the New York Academy of Science*, 280, pp. 173-216.

Chomsky, N. and Caplan, D. (1980), 'Linguistic Perspectives on Language Development', in D. Caplan (ed.) (1980), pp. 97-105.

Chomsky, Noam (1980), 'Rules and Representations', *The Behavioural and Brain Sciences*, 3, pp. 1-61.

Chung, S. (1977), 'On the gradual nature of syntactic change', in C. N. Li (ed.), *Mechanisms of Syntactic Change*, Austin: University of Texas Press, pp. 3-55.

Clarke, A. M. and Clarke, A. D. B. (eds) (1976),*Early Experience: Myth and Evidence*, London: Open Books.

Clarke, B. (1970), 'Selective Constraints on Amino Acid Substitutions During Evolution of Proteins', *Nature*, 228, pp. 159-60.

'Cognition and Consciousness in Non-Human Species: Open Peer Commentary and Authors' Responses' (1978), *The Behavioural and Brain Sciences*, 4, pp. 555-629.

Cooke, M. and McHenry, R. (1978), *Sexual Attraction*, Oxford: Pergamon Press.

Cowey, A. and Weiskrantz, L. (1975), 'Demonstration of Cross-Modal Matching in Rhesus Monkeys, *Macaca mulatta*', *Neuropsychologia*, 13, pp. 117-20.

Crick, F. H. C. (1958), 'On Protein Synthesis', *Symp. Soc. Exp. Biol.*, 12, pp. 138-163.

D'Aquili, E. G. (1978), 'Neurobiological Bases of Myth and Concepts of Deity', *Zygon*, 13, pp. 257-75.

Darlington, C. D. (1969), *The Evolution of Man and Society*, London: George Allen & Unwin.

Darlington, C. D. (1978), *The Little Universe of Man*, London: George Allen & Unwin.

Darwin, C. (1859), *On The Origin of Species*, London.

Darwin, C. (1874), *The Descent of Man and Selection in Relation to Sex*, New York: Hurst.

Darwin, C. and Wallace, A. H. (1958), *Evolution by Natural Selection*, Cambridge University Press.

Dato, D. P. (1975), *Georgetown Roundtable on Languages and Linguistics: Developmental Psycholinguistics*, Washington DC: Georgetown University Press.

Davenport, R. K. (1976), 'Cross-Modal Perception in Apes', in Harnad *et al.* (eds) (1976), pp. 143-9.

Davidson, D. (1967), 'Causal Relations', *Journal of Philosophy*, 64, pp. 691-703.

Davidson, D. (1973), 'Radical Interpretation', *Dialectica*, XXVII, pp. 313-28.

Davidson, D. (1975), 'Thought and Talk', in Guttenplan (ed.) (1975), pp. 7-23.

Davidson, D. (1979), 'Moods and Performances', in Margalit (ed.) (1979), pp. 9-20.

Davis, J. W. F. and O'Donald, P. (1976), 'Sexual Selection for a Handicap: Critical Analysis of Zahavi's Model', *Journal of Theoretical Biology*, 57, pp. 345-54.

Dawkins, R. (1976), *The Selfish Gene*, New York: Oxford University Press.

Dawkins, R. (1979), 'Twelve Misunderstandings of Kin Selection', *Zeitschrift für Tierpsychologie*, 51, pp. 184-200.

Dawkins, R. (1981), 'In Defence of Selfish Genes', *Philosophy*, 56, pp. 556-73.

Dennett, D. C. (1981), 'What Is It Like To Be Me?', *New Scientist*, 92, pp. 806-9.

Dewey, J. (1910), *The Influence of Darwin on Philosophy and Other Essays in Contemporary Thought*, New York.

Dickemann, M. (1979), 'Female Infanticide, Reproductive Strategies, and Social Stratification: A Preliminary Model', in Chagnon and Irons (eds) (1979).

Dingwall, W.O. (1975), 'The Species-Specificity of Speech', in D.P. Dato (ed.) (1975).

Dingwall, W. O. (1979), 'The Evolution of Human Communication Systems', H. Whitaker and H. A. Whitaker (eds), *Studies in Neurolinguistics*, vol. 4, New York: Academic Press.

Ditfurth, H. V. (1976), *Der Geist fiel nicht vom Himmel*, Hamburg Hoffman & Campe.

Dobzhansky, T. (1970), *Genetics of the Evolutionary Process*, New York: Columbia University Press.

Dobzhansky, T. and Spassky, B. (1976a), 'An Experiment on Migration and Simultaneous Selection for Several Traits in *Drosophila pseudo obscura*', *Genetics*, 55, pp. 723-34.

Dobzhansky, T. and Spassky, B. (1976b), 'Effects of Selection and Migration on the Geotactic and Phototactic Behaviour of Drosophila', *Proceedings of the Royal Society*, 168, pp. 27-47.

Dobzhansky, T., Ayala, F. J., Stebbins, R. C. and Valentine, J. W. (1977), *Evolution*, San Francisco: Will Freeman & Co.

Drach, M. (1981), 'The Creative Aspect of Chomsky's Use of the Notion of Creativity', *The Philosophical Review*, XC, no. 1, pp. 44-65.

Du Brul, E. L. (1979), '(Notes and Discussions) Poor Old Neanderthal Man: Responses to a Response', *Brain and Language*, 8, pp. 251-7.

Durkheim, E. (1938), *The Rules of Sociological Method*, Chicago: Free Press.

Eccles, H. (1977), 'Evolution of the Brain in Relation to the Development of the Self-Conscious Minds', *Annals of the New York Academy of Sciences*, 292, pp. 161-79.

Ehrman, L. (1972), 'Genetics and Sexual Selection', in C. Campbell (ed.) (1972).

Eibl-Eibesfeldt, I. (1967), *Grundriss der Vergleichenden Verhaltensforschung*, Munich: R. Piper & Co.

Eibl-Eibesfeld (1973), *Der Vorprogrammierte Mensch*, Vienna: Fritz Molden.

Eigen, M. (1971), 'Self-Organisation of Matter and the Evolution of Biological Macromolecules', *Naturwissenschaften*, 58, pp. 465-523.

Eigen, M. (1978), 'The Hypercycle. A Principle of Natural Self-Organisation', *Naturwissenschaften*, 65, pp. 341-69.

Eigen, M., Gardiner, W., Schuster, P., and Winkler-Oswatitsch, R. (1981), 'Ursprung der Genetischen Information', *Spektrum der Wissenschaft*, pp. 37-56, June.

Eigen, M. and Winkler-Oswatitsch, R. (1981), 'Transfer-RNA, an Early Gene?', *Naturwissenschaften*, 68, pp. 282-92.

Eigen, M. and Winkler, R. (1975), *Das Speil* , Munich: R. Piper & Co.

Eisner, T. and Wilson, E. O. (eds) (1977), *The Insects*, readings from *Scientific American*, San Francisco: W. H. Freeman.

Eldredge, N. and Cracraft, J. (1980), *Phylogenetic Pattern and the Evolutionary Process*, New York: Columbia University Press.

Eldredge, N. and Gould, S.J. (1972), 'Punctuated Equilibria: an alternative to phyletic gradualism', in T.M. Schopf (ed.), *Models in Paleobiology*, San Francisco: Freeman, Cooper.

Epstein, R., Lanza, R. P. and Skinner, B. F. (1980), 'Symbolic Communication Between Two Pigeons', *Science*, 207, pp. 543-5, February.

Evans, H, St. B. (1982) *The Psychology of Deductive Reasoning*, London: Routledge & Kegan Paul.

Eysenck, H. J. (ed.) (1980), *A Model for Personality*, Berlin: Springer.

Falconer, D. S. (1960), *Introduction to Quantitative Genetics*, Edinburgh: Oliver & Boyd.

Feldman, M. W. and Lewontin, R. C. (1975), 'The Heritability Hang-Up', *Science*, 90, pp. 1163-8.

Feyerabend, P. (1975), *Against Method. Outline of an Anarchistic Theory of Knowledge*, London: New Left Books.

Field, H. H. 'Mental Representation', *Erkenntnis*, 13, pp. 9-61.

Fisher, R. A. (1930), *The Genetical Theory of Natural Selection*, Oxford: Clarendon Press.

Flew, A. (1967), *Evolutionary Ethics*, London: Macmillan.

Føllesdal, D. (1979), 'Hermeneutics and the Hypothetico-Deductive Method', *Dialectica*, 33, pp. 319-36.

Fourcher, L. A. (1978), 'A View of Subjectivity in the Evolution of Communicative Behaviour', *Journal of Social Biological Structure*, 1, pp. 387-400.

Fouts, R., *et al.* (1976), 'Translation of Signed Responses in American Sign Language from Vocal English Stimuli to Physical Object Stimuli by a Chimpanzee', *Learning and Motivation*, 7, 458-75.

Fox, S. W. (1973), 'Origin of the Cell: Experiments and Premises', *Naturwissenschaften*, 60, pp. 359-68.

Franck, D. (1976), 'Gibt es Wirklich Keinen Agressionstrieb?', *Umschau*, 76, pp. 309-14.

Franklin, I. and Lewontin, R. C. (1970), 'Is the Gene the Unit of Selection?', *Genetics*, 65, pp. 707-34.

Freedman, D. G. (1980), *Human Sociobiology*, New York: Macmillan.

Frege, G. (1879), *Begriffsschrift, eine der Arithmetischen Nachgebildete Formelsprache des Reinen Denkens*, Halle an der Saale.

Frye, N. (1981), 'The Bridge of Language', *Science*, 212, pp. 127-32, April.

Fuller, J. L. and Thompson, W. R. (1978), *Foundations of Behaviour Genetics*, Saint Louis: The C. V. Mosby Co.

Gallup, G. G. (1979), 'Self-Awareness in Primates', *American Scientist*, 67, pp. 417-21.

Gazdar, G. (1981), 'Phrase Structure Grammar', in G. K. Pullum and P. Jacobson (eds), *The Nature of Syntactic Representation*, Dordrecht: Reidel.

Gazzaniga, M.S. (1970), *The Bisected Brain*, New York: Appleton-Century-Crofts.

Geist, V. (1978), *Life Strategies, Human Evolution, Environmental Design, Toward a Biological Theory of Health*, New York: Springer.

Geschwind, N. (1974), *Language and the Brain*, Dordrecht: Reidel.

Ghiselin, M. T. (1974), *The Economy of Nature and the Evolution of Sex*, University of California, Berkeley.

Gierer, A. (1974), 'Hydra as a Model for the Development of Biological Form', *Scientific American*, 231, pp. 44-54.

Goldin-Meadow, S. J. (1975), 'The Representation of Semantic Relations in a Manual Language Created by Deaf Children of Hearing Parents: a Language You Can't Dismiss Out of Hand', PhD thesis, Department of Psychology, University of Pennsylvania, quoted in D. MacNeill (1979).

Goldschmidt, R. (1940), *The Material Basis of Evolution*, New Haven, Conn.: Yale University Press.

Goodall, J. (1979), 'Life and Death at Gombe', *National Geographic*, pp. 592-621.

Gorczynski, R. M. and Steele, E. J. (1981), 'Simultaneous Yet Independent Inheritance of Somatically Acquired Tolerance on Two Distinct H-2 Antigenic Halotype Determinants in Mice', *Nature*, 289, pp. 678-81.

Gould, S. J. (1977), *Ontogeny and Phylogeny*, Cambridge, Mass.: Harvard University Press.

Gould, S. J. and Eldredge, N. (1977), 'Punctuated Equilibria: The Tempo and Mode of Evolution Reconsidered', *Paleobiology*, pp. 115-51.

Gould, S. J. and Lewontin, R. C. (1979), 'The Spandrels of San Marco and the Panglossian Paradigm: a Critique of the Adaptionist Programme', *Proc. R. Soc. Lond. B. Biol.*, 205, pp. 581-8.

Greenberg, J. H. (1963), *Universals of Language*, Cambridge, Mass.: MIT Press.
Gribbin, J. and Cherfas, J. (1981), 'Descent of Man – Or Ascent of Ape?', *New Scientist*, 91, pp. 592-5.
Grice, H. P. (1957), 'Meaning', *Philosophical Review*, 66, pp. 377-88.
Grice, H. P. (1969), 'Utterer's Meaning and Intentions', *Philosophical Review*, 78, pp. 147-77.
Guttenplan, S. (ed.) (1975), *Mind and Language*, London: Oxford University Press.
Hacking, I. (1975), *The Emergence of Probability: a Philosophical Study of Early Ideas About Probability, Induction and Statistical Inference*, Cambridge University Press.
Halstead, B. (1980), 'Popper: Good Philosophy, Bad Science?', *New Scientist*, 90, pp. 215-17.
Halvorson, H. O. and Van Holde, K. E. (1981), 'The Origins of Life and Evolution', New York: Alan R. Liss.
Hamilton, W. D. (1964), 'The Genetical Theory of Social Behaviour', I and II, *Journal of Theoretical Biology*, 7, pp. 1-52.
Hardin, G. (1968), 'The Tragedy of the Commons', *Science*, 162, pp. 1243-8.
Hardin, G. (1977), *The Limits of Altruism*, Bloomington: Indiana University Press.
Harlow, H. F. and Harlow, M. K. (1962), 'Social Deprivation in Monkeys', *Scientific American*, 207, pp. 136-46.
Harnad, S. R., Steklis, H. D. and Lancaster, J. (eds) (1976), *Origins and Evolution of Language and Speech. Annals of the New York Academy of Sciences*, vol. 280.
Haugeland, J. (1978), 'The Nature and Plausibility of Cognitivism', *The Behavioural and Brain Sciences*, 2, pp. 215-60.
Hayek, F. A. (1978), *Three Sources of Human Values*, London School of Economics.
Hayes, C. (1952), *The Ape in Our House*, London: Gollancz.
Heiligenberg, W. and Kramer, U. (1972), 'Aggressiveness as a Function of External Stimulation', *Journal of Comparative Physiology*, 77, pp. 332-40.
Hellman, G. and Thompson, F. (1975), 'Physicalism: Ontology, Determination and Reduction', *Journal of Philosophy*, 72, pp. 551-64.
Hempel, C. G. (1962), 'Explanation in Science and in History', from R. G. Colodny (ed.), *Frontiers of Science and Philosophy*, London: Allen & Unwin and University of Pittsburgh Press, pp. 7-33.
Hempel, C. G. (1965), *Aspects of Scientific Explanation*, New York: The Free Press.
Hempel, C.G. (1966), *Philosophy of Natural Sceince*, Englewood Cliffs, N.J.: Prentice-Hall.
Herrnstein, R. (1973), *IQ in the Meritocracy*, Harmondsworth: Penguin.
Hess, E. H. (1956), 'Space Perception in the Chicken', *Scientific American*, 195, pp. 71-80.
Hewes, G. (1973), 'Primate Communication and the Gestural Origin of

Language', *Current Anthropology*, 14, pp. 5-24.
Hewes, G. (1973), 'An Explicit Formulation of the Relationship Between Tool-Using, Tool-Making and the Emergence of Language', *Visible Language*, 7 (2), pp. 191-227.
Hewes, G. W. (1977), 'Language Origin Theories', in Rumbaugh, (ed.) (1977), pp. 3-53.
Hinde, R. A. (1974), *Biological Bases of Human Social Behaviour*, New York: McGraw-Hill.
Hirshleifer, J. (1977), 'Economics from a Biological Viewpoint', *Journal of Law and Economics*, 20, pp. 1-52.
Ho, M. W. and Saunders, P. T. (1979), 'Beyond Neo-Darwinism – An Epigenetic Approach to Evolution', *J. Theor. Bio.*, 78, pp. 573-91.
Hockett, C. (1960), 'The Origin of Speech', *Scientific American*, 203, pp. 88-96.
Holden, C. (1979), 'Paul MacLean and the Triune Brain', *Science*, 204, pp. 1066-8.
Holden, C. (1980), 'Twins Reunited', *Science*, 80, pp. 55-9.
Holloway, R. L. (1976), 'Paleoneurological Evidence for Language Origin', in Harnad *et al.*, (eds) (1976), pp. 330-48.
Hopcroft, J.E. and Ullman, J.D. (1969), *Formal Languages and Their Relation to Automata*, London: Addison-Wesley.
Hull, D. (1981), 'Units of Evolution: a Metaphysical Essay', in U. J. Jensen and R. Harné (eds), *The Philosophy of Evolution*, Brighton: Harvester.
Hume, D. (1977), *A Treatise of Human Nature*, ed. L. A. Selby-Bigge, London: Oxford University Press.
Humphrey, N. K. (1979), 'Nature's Psychologists', in B. D. Josephenson and V. S. Ramachandra, (eds), *Consciousness and the Physical World*, Oxford: Pergamon Press, pp. 57-76.
Huxley, J. S. (1963), *Evolution, the Modern Synthesis*, New York: Harper.
Huxley, T. H. (1947), Prolegomena to 'Evolution and Ethics' (1984), in T. H. Huxley and Julian Huxley, *Evolution and Ethics 1893-1943*, London: Pilot Press.
Ingram, D. (1975), 'If and When Transformations are Acquired by Children', in D. P. Dato (ed.), *Georgetown Roundtable on Languages and Linguistics: Developmental Psycholinguistics*, St. Louis, MO: University of Washington Press.
Inhelder, B. and Piaget, J. (1964), *The Early Growth of Logic*, New York: W. W. Norton.
Irons, W. (1979), 'Cultural and Biological Success', in Chagnon and Irons (eds) (1979).
Isaac, G. L. (1976), 'Stages of Cultural Elaboration in the Pleistocene: Possible Archaeological Indicators of the Development of Language Capabilities', in Harnad *et al.*, (eds) (1976), pp. 275-88.
Jacobs, J. (1981), 'How Heritable is Innate Behaviour?', *Zeitschrift für Tierpsychologie*, 55, pp. 1-18.
Janssen, T., Kok, G. and Meertens, L. (1977), 'On Restrictions on

Transformational Grammars Reducing the Generative Power', *Linguistics and Philosophy*, 1, pp. 111-18.
Jarvis, M. J. and Ettlinger, G. (1977), 'Cross-Model Recognition in Chimpanzees and Monkeys', *Neuropsychologia*, 15, pp. 499-506.
Jaynes, J. (1976), *The Origin of Consciousness in the Breakdown of the Bicameral Mind*, Boston: Houghton Mifflin.
Jaynes, J. (1976), 'The Evolution of Language in the Late Pleistocene', in Harnad *et al.*, (eds) (1976), pp. 312-25.
Jensen, A. R. (1972), *Genetics and Education*, London: Methuen.
Jones, J. S. (1980), 'How Much Genetic Variation?', *Nature*, 288, pp. 10-11.
Jones, J. S. (1981), 'Models of Speciation – The Evidence from *Drosophila*', *Nature*, 289, pp. 743-4.
Kant, I. (1785) *Grundlegrung zur Metaphysik der Sitten*, Rig: Hartknoch.
Kant, I. (1787), *Critique of Pure Reason*, trans. N. Kemp Smith, London: Macmillan, 2nd edn., 1929.
Kettlewell, H. B. D. (1961), 'The Phenomenon of Industrial Melanism in the Lepidoptera', *Annual Review of Entomology*, 6, pp. 245-62.
Kiefer, J. (1979), 'Physik und Leben', in B. Kanitscheider (ed.), *Materie-Leben-Geist*, Berlin: Duncker & Humbolt.
Kimura, M. and Ohta, T. (1971), *Theoretical Aspects of Population Genetics*, Princeton University Press.
King, M. C. and Wilson, A. C. (1975), 'Evolution at Two Levels in Humans and Chimpanzees', *Science*, 188, pp. 107-16.
Knight, G. R., *et al.* (1956), 'Selection for Sexual Isolation within a Species', *Evolution*, 10, pp. 14-22.
Koestler, A. (1971), *The Case of the Midwife Toad*, London: Hutchinson.
Kørner, S. (1955), *Kant*, Harmondsworth: Penguin.
Kuhn, H. and Waser, J. (1981), 'Molekulare Selbstorganisation und Ursprung des Lebens', *Angewandte Chemie*, 93, pp. 495-515.
Kuhn, T. S. (1970), *The Structure of Scientific Revolutions*, 2nd edn, University of Chicago Press.
Kunkel, J. (1977), 'Sociobiology vs Bio-Sociology', *American Sociologist*, 12, pp. 69-73.
Kuo, Z.Y. (1970) 'The Need for Co-Ordinated Efforts in Developmental Studies', in L. R. Aronson *et al.* (eds), *Development and Evolution of Behaviour*, San Francisco: W. H. Freeman.
Lamarck, J. B. de (1914), *Zoological Philosophy*, London: Macmillan.
Lamendella, J. T. (1976), 'Relations Between the Ontogeny and Phylogeny of Language', *Annals of the New York Academy of Sciences*, 280, pp. 396-412.
Laughlin, C. D. and D'Aquili, E. G. (1974), *Biogenetic Structuralism*, New York: Columbia University Press.
Laughlin, C. D. and D'Aquili, E. G. (1975), 'The Biopsychological Determinants of Religious Ritual Behaviour', *Zygon*, 10, pp. 32-58.
Leakey, R. E. (1981a), 'The Birth of Language', *Listener* (28 May 1981).
Leakey, R. E. (1981b), 'The Making of Mankind IV', *Listener* (28 May

1981), pp. 700-2.

Leakey, R. E. (1981c), *The Making of Mankind*, New York: E. P. Dutton.

Lederberg, J. and Lederberg, E. M. (1952), 'Replica Plating and Indirect Selection of Bacterial Mutants', *J. Bacteriol.*, 63, pp. 399.

Lehrman, D. S. (1970), 'Semantic and Conceptual Issues in the Nature-Nurture Problem', in L. R. Aronson, E. Tobach, D. S. Lehrman and J. S. Rosenblatt (eds), *Development and Evolution of Behaviour*, San Francisco: W.H. Freeman & Co.

Lenneberg, E. H. (1966), 'The Natural History of Language', in Smith and Miller, (eds) (1966), pp. 219-52.

Lerner, I.M. (1968), *Heredity, Evolution and Society*, San Francisco: W. H. Freeman & Co.

Lewin, R. (1980), 'Is Your Brain Really Necessary?', *Science*, 210, pp. 1232-4.

Lewin, R. (1981), 'Do Jumping Genes Make Evolutionary Leaps?', *Science*, 213, pp. 634-6.

Lewis, D.K. (1969), *Convention*, Cambridge, Mass.: Harvard University Press.

Lewontin, R. C. (1969), 'The Bases of Conflict in Biological Explanation', *Journal of the History of Biology*, 2, pp. 35-45.

Lewontin, R. C. (1970), 'The Units of Selection', *Annual Review of Ecology* and *Systematics*, 1, pp. 1-18.

Lewontin, R. C. (1974), *The Genetic Basis of Evolutionary Change*, New York: Columbia University Press.

Lewontin, R. C. (1978), 'Single- and Multiple-Locus Measures of Genetic Distance Between Groups', *American Naturalist*, 112, pp. 1138-9.

Li, C. (1977), *Mechanisms of Syntactic Change*, Austin, University of Texas Press.

Llinas, R. R. (1981), 'Cerebellar Modelling', *Nature*, 291, pp. 279-80, 28 May.

Loehlin, J.C. and Nichols, R.C. (1976), *Heredity, Environment, and Personality*, Austin, University of Texas Press.

Lorenz, K. (1941), 'Kants Lehre vom Apriorischen im Lichte gegenwärtiger Biologie', *Blätter für Deutsche Philosophie*, 15, pp. 94-125.

Lorenz, K. (1943), 'Die angeborenen Formen möglicher Erfahrung', *Zeitschrift für Tierpsychologie*, 5, pp. 235-409.

Lorenz, K. (1963), *Das Sogenannte Böse*, Vienna: Borotha-Schoeler.

Lorenz, K. (1965), *Evolution and Modification of Behaviour*, University of Chicago Press.

Lorenz, K. (1970), 'The Enmity Between Generations and Its Probable Ethological Causes', in A. Tiselius and S. Nilsson (eds), *The Place of Value in a World of Facts*, New York: Wiley.

Lorenz, K. (1973), *Die Rückseite des Spiegels*, Munich: Piper.

Lorenz, K. (1966), *On Aggression*, New York: Harcourt Brace Jovanovich.

Losco, J. (1981), 'Ultimate vs Proximate Explanation: Explanatory Modes in Sociobiology and the Social Sciences', *Journal of Social and Biological Structures*, 4, pp. 329-46.

Lovtrup, S. (1976), 'On the Falsifiability of Neo-Darwinism', *Evol. Theory.*, 1, pp. 267-83.

Lucas, J.R. (1964), 'Minds, Machines and Gödel', in A.R. Anderson (ed.), *Minds and Machines*, Englewood Cliffs, N.J.: Prentice-Hall.

Lumsden, C.J. and Wilson, E.O. (1981), *Genes, Mind and Culture: The Coevolutionary Process*, Cambridge, Mass.: Harvard University Press.

Lyons, J. (1972), Review of *Chomsky*, (Modern Masters, (ed.) Frank Kermode, New York, Viking Press, 1970), in *Language*, vol. 48, no. 2, pp. 416-27.

McCulloch, W. S. and Pitts, W. (1943), 'A Logical Calculus of the Ideas Immanent in Nervous Activity', *Bulletin of Mathematical Biophysics*, 5, pp. 115-33.

McDowell, J. (1980), 'Meaning, Communication and Knowledge', in van Straaten (ed.) (1980), pp. 117-39.

McGinn, C. (1979), 'Evolution, Animals and the Basis of Morality', *Inquiry*, 22, pp. 81-99.

McGonigle, B. (1980), 'Sign, Symbol and Syntax in the Language of Apes', *Nature*, 286, pp. 761-2, 21 August.

McGonigle, Brendan and Chalmers, Margaret (1977), 'Are Monkeys Logical?', *Nature*, 267, pp. 694-6, 23 June.

McGrew, W. C. and Tutin, C. E. G. (1978), 'Evidence for a Social Custom in Wild Chimpanzees?', *Man* (NS), 13, pp. 234-251.

McGrew, W. C., Tutin, C. E. G. and Baldwin, P. J. (1979), 'Chimpanzees, Tools, and Termites: Cross-Cultural Comparisons of Senegal, Tanzania, and Rio Muni', *Man* (NS), 14, pp. 185-214.

Mackie, J. L. (1981), 'Genes and Egoism', *Philosophy*, 56, pp. 553-5.

McLaren, A., *et al.* (1981), 'Immune Reactivity of Progeny of Tetraparental Male Mice', *Nature*, 290, pp. 513-14.

MacNeill, D. (1979), 'Language Origins', in von Cranach *et al.* (eds) (1979), pp. 715-28.

Malthus, T. R. (1798), *An Essay on the Principle of Population*, ed. A. Flew, Harmondsworth, Penguin.

Margalit, A. (ed.) (1979), *Meaning and Use*, Dordrecht: Reidel.

Markl, H. (ed.) (1980), *Evolution of Social Behaviour: Hypotheses and Empirical Tests*, Weinheim: Verlag Chemie.

Marshack, A. (1976), 'Some Implications of the Paleolithic Symbolic Evidence for the Origin of Language', in Harnad *et al.* (eds) (1976), pp. 289-311.

Marshall, J. C. (1980), 'On the Biology of Language Acquisitions', in Caplan (ed.) (1980), pp. 106-48.

Marx, J. L. (1980), 'Ape-Language Controversy Flares Up', *Science*, 207, pp. 1330-3.

Mather, K. and Jinks, J.L. (1971), *Biometrical Genetics*, London: Chapman and Hall.

Matthews, R.J. 'Are the Grammatical Sentences of a Language a Recursive Set?', *Synthese*, 40, pp. 209-24.

Maynard Smith, J. (1977a), 'The Limitations of Evolutionary Theory', in R. Duncan and M. Weston-Smith (eds), *The Encyclopedia of Ignorance*, Oxford, Pergamon Press.

Maynard Smith, J. (1977b), 'Why the Genome does not Congeal', *Nature*, 268, pp. 693-6.

Maynard Smith, J. (1976a), 'Group Selection', *Quarterly Review of Biology*, 51, pp. 277-83.

Maynard Smith, J. (1976b), 'Sexual Selection and the Handicap Principle', *Journal of Theoretical Biology*, 57, pp. 239-42.

Maynard Smith, J. and Parker, G. A. (1976), 'The Logic of Assymetric Contests', *Animal Behaviour*, 24, pp. 159-75.

Maynard Smith, J. and Price, G. R. (1973), 'The Logic of Animal Conflict', *Nature*, 246, pp. 15-18.

Mayr, E. (1963), *Animal Species and Evolution*, Cambridge, Belknap Press.

Mayr. E. (1972), 'The Nature of the Darwinian Revolution', *Science*, 176, pp. 981-9.

Medawar, P., Simpson, E., Chandler, P., Fierz, W., Brent, L. and Rayfield, L.S. (1982), 'Further Studies on Supposed Lamarkian Inheritance of Immunological-Tolerance', *Nature*, 295, p. 242.

Medawar, P. (1961) 'Review of *The Phenomenon of Man*', *Mind*, LXX, pp. 99-106.

Mellor, D.H. (1980), 'Consciousness and degrees of belief' in *Prospects for Pragmatism: Essays in Memory of F.P. Ramsey*, Cambridge University Press.

Michod, R.E. and Hamilton, W.D. (1980), 'Coefficients of Relatedness in Sociobiology', *Nature*, 288, pp. 694-7.

Midgley, M. (1979), 'Gene Juggling', *Philosophy*, 54, pp. 439-58.

Midgley, M. (1980), 'Rival fatalisms: The Hollowness of the Sociobiology Debate, in A. Montagu (ed.), *Sociobiology Examined*, New York: Oxford University Press, pp. 15-38.

Mill, J. S. (1863), *Utilitarianism*, London.

Miller, S. and Orgel, L. E. (1973), *The Origins of Life on Earth*, Englewood Cliffs, N. J.: Prentice-Hall.

Mills, D.R., Peterson, R.L. and Spiegelman, S. (1967), 'An extracellular Darwinian experiment with a self-duplicating nucleic acid molecule', *Proceedings of the National Academy of Science US*, 58, p. 217.

Mises, R. von (1961), *Probability, Statistics and Truth*, 2nd revised edn, London: Allen & Unwin.

Mitton, J. B. (1977), 'Genetic Differentiation of Races of Man as Judged by Single-Locus and Multi-Locus Analyses', *American Naturalist*, 111, pp. 203-12.

Monod, J. (1971), *Chance and Necessity*, New York: Knopf.

Montague, R. (1974), *Formal Philosophy*, New Haven, Conn.: Yale University Press.

Moore, G. E. (1959), 'Proof of an External World', *Philosophical Papers*, London: Allen & Unwin, pp. 127-50.

Mortiz, K. and Roth, G. 'Complexity of Germ Line and Somatic DNA in Ascari', *Nature*, 259, pp. 55-7.

Mundinger, P.C. (1980), 'Animal Cultures and a General Theory of Evolution', *Ethology and Sociobiology*, 1, pp. 183-223.

Murray, R. D. (1980), 'The Evolution and Functional Significance of Incest Avoidance', *Journal of Human Evolution*, 9, pp. 173-8.

Neel, J.V. (1981), 'The Major Ethnic Groups: Diversity in the Midst of Similarity', *American Naturalist*, 117, pp. 83-7.

O'Donald, P. (1962) 'The Theory of Sexual Selection', *Heredity*, 17, pp. 541-52.

Olivier, G. (1979), 'Causes of Biological Differences Among Social Classes', *Journal of Human Evolution*, 8, pp. 813-16.

Orgel, L. E. and Crick, F. H. C. 'Selfish DNA: the Ultimate Parasite', *Nature*, 284, pp. 604-7.

Parker, C. E. (1978), 'Opportunism and the Rise of Intelligence', *Journal of Human Evolution*, 7, pp. 597-608.

Parker, S.T. and Gibson, K.R. (1979), 'A developmental model of the evolution of language and intelligence in early hominids', *The Behavioural and Brain Sciences*, 2, pp. 367-407.

Passingham, R. E. and Ettlinger, G. (1974), 'A Comparison of Cortical Functions in Man and Other Primates', *International Review of Neurobiology*, 16, pp. 233-99.

Peacocke, A. R. (1983), *An Introduction to the Physical Chemistry of Biological Organisation*, London: Oxford University Press.

Peirce, C. S. (1958), *Collected Papers*, ed. A Burks, Cambridge, Mass.: Harvard University Press.

Peters, P. S. and Ritchie, R. W. (1971), 'On Restricting the Base Component of Transformational Grammar', *Information and Control*, 18, pp. 483-501.

Peters, P. S. (1973), 'On the Generative Power of Transformational Grammars', *Information Sciences*, 6, pp. 49-83.

Petitto, L. A. and Seidenberg, M. S. (1979), 'On the Evidence for Linguistic Abilities in Signing Apes', *Brain and Language*, 8, pp. 162-83.

Piaget, J. (1952), *The Origins of Intelligence in Children*, New York: International University Press.

Pinker, S. (1979), 'Formal Models of Language Learning', *Cognition*, 7, pp. 217-83.

Popper, K. R. (1957), *The Poverty of Historicism*, Boston: Beacon Press.

Popper, K. R. (1959), *The Logic of Scientific Discovery*, New York: Basic Books.

Popper, K. R. (1966), *Of Clouds and Clocks: an Approach to the Problem of Rationality and the Freedom of Man*, St Louis: Washington University Press.

Popper, K. R. (1969), *Conjectures and Refutations*, London: Routledge & Kegan Paul.

Popper, K. R. (1974), *The Unended Quest*, London: Collins.

Post, R.H. (1962), 'Population Differences in Visual Acuity: A Review with Speculative Notes on Selection Relaxation', *Eugenics Quarterly*, 9, pp. 189-212.

Postlethwaite, J.H. and Schneiderman, H.A. (1973), 'Developmental Genetics of *Drosophila* Imaginal Discs', *Annual Review of Genetics*, 7, pp. 381-433.

Premack, D. (1976), *Intelligence in Apes and Man*, New Jersey: Erlbaum.

Premack, D. and Woodruff, G. (1978), 'Chimpanzee Problem-Solving: A Test for Comprehension', *Science*, 202, pp. 532-5, 3 November.

Premack, D. and Woodruff, G. (1978), 'Does the Chimpanzee Have a Theory of Mind?', *The Behavioural and Brain Sciences*, 4, pp. 515-26.

Premack, D. and Woodruff, G. (1979), 'Chimpanzee Problem Comprehension: Insufficient Evidence', *Science*, 306, pp. 1201-2, 7 December.

Premack, D. and Woodruff, G. (1979), 'Intentional Communication in the Chimpanzee: The Development of Deception', *Cognition*, 7, pp. 333-62.

Prigogine, I. (1979), *Vom Sein zum Werden*, Munich: Piper.

Prigogine, I. (1980), *From Being to Becoming: Time and Complexity in the Physical Sciences*, San Francisco: W. H. Freeman & Co.

Prigogine, I., and Stengers, I. (1981), *Dialog mit der Natur*, Munich: Piper Verlag.

Puccetti, R. and Dykes, R. W. (1978), 'Sensory Cortex and the Mind-Brain Problem', *The Behavioural and Brain Sciences*, 3, pp. 337-5.

Pugh, G.E. (1977), *The Biological Origin of Human Values*, London: Routledge & Kegan Paul.

Pulliam, H. R. and Dunford, C. (1980), *Programmed to Learn*, New York: Columbia University Press.

Pylyshyn, Z. W. (1974), 'Minds, Machines and Phenomenology: Some Reflections on Dreyfus "What Computers Can't Do" ', *Cognition*, vol. no. 3, 1, pp. 57-77.

Quine, W. V. O. (1953), 'Two Dogmas of Empiricism' in *From a Logical Point of View*, Cambridge, Mass.: Harvard University Press.

Quine, W. V. O. (1969), 'Epistemology Naturalised', in *Ontological Relativity and other Essays*, New York: Columbia University Press.

Quinton, A. (1965), 'Ethics and the Theory of Evolution', in I. T. Ramsey (ed.): *Biology and Personality*, Oxford: Blackwell, pp. 197-231.

Radcliffe-Brown, A. R. (1952), *Structure and Function in Primitive Society*, Chicago, Free Press.

Rasa, O. A. E. (1971), 'Appetence for Aggression in Juvenile Damsel Fish', *Zeitschrfit für Tierpsychologie*, Suppl. 7, 70, pp. 1-68.

Reader, J. (1981), 'Whatever Happened to Zinjanthropus', *New Scientist*, 92, pp. 802-5.

Reed, E. S. (1981), 'The Lawfulness of Natural Selection', *American Naruralist*, 118, pp. 61-71.

Rensch, B. (1960), *Evolution Above the Species Level*, New York: Columbia University Press.
Ridley, M. (1981), 'Who Doubts Evolution', *New Scientist*, 91, pp. 830-2.
Riedl, R. (1976), *Die Strategie der Genesis*, Munich: Piper.
Riedl, R. (1977), 'A Systems-Analytic Approach to Macro-Evolutionary Phenomena', *Quarterly Review of Biology*, 52, pp. 251-70.
Riedl, R. (1979), *Biologie der Erkenntnis*, Berlin: Paul Parey.
Riedl, R. (1980), *Biologie der Erkenntnis*, Berlin: Paul Parey.
Root-Bernstein, R. (1981), Letter *Science*, 212, pp. 1446-8.
Routley, R. (1981), 'Alleged Problems in Attributing Beliefs and Intentionality to Animals', *Inquiry*, 24, pp. 385-41.
Royce, J. R. and Mos, L. P. (eds) (1979), *Theoretical Advances in Behaviour Genetics*, Alphen aan den Rijn, Sijthoff & Nordhoff Int. Publishers.
Rumbaugh, D. M. (ed.) (1977), *Language Learning by a Chimpanzee*, New York: Academic Press.
Ruse, M. (1979a), *The Darwinian Revolution*, University of Chicago Press.
Ruse, M. (1979b), *Sociobiology: Sense or Nonsense*, Dordrecht: Reidel.
Ruyle, E. E. (1977), 'The Adaptive Significance of Cultural Behaviour: Comments and Reply', *Human Ecology*, 5, pp. 53-5.
Sahlins, M. D. (1976), *The Use and Abuse of Biology: An Anthropological Critique of Sociobiology*, Ann Arbor: University of Michigan Press.
Sampson, G. (1978), 'Linguistic Universals as Evidence for Empiricism', *Journal of Linguistics*, 14, pp. 129-375.
Sampson, G. (1980), 'Popperian Language-Acquisition Undefeated', *British Journal for Philosophy of Science*, 31, pp. 63-7.
Sands, S. F. and Wright, A. A. (1980), 'Primate Memory: Retention of Serial List Items by a Rhesus Monkey', *Science*, 209, pp. 938-40, 22 August.
Sanides, F. (1975), 'Comparative Neurology of the Temporal Lobe in Primates Including Man with Reference to Speech', *Brain and Language*, 2, pp. 396-419.
Saunders, P.T. and Ho, M.W. (1976), 'On the Increase in Complexity in Evolution', *Journal of Theoretical Biology*, 63, pp. 375-84.
Savage-Rumbaugh, E. S., Rumbaugh, D. M. and Boysen, S. (1978), 'Linguistically Mediated Tool Use and Exchange by Chimpanzees (*Pan troglodytes*)', *The Behavioural and Brain Sciences*, 4, pp. 539-54.
Savage-Rumbaugh, E. S., Rumbaugh, D. M. and Boysen, S. (1980), 'Do Apes Use Language?', *American Scientist*, 68, pp. 49-61, January/February.
Scheinfeld, A. (1965), *Your Heredity and Environment*, Philadelphia: Lippincott.
Schilcher, F. von (1978), 'Verhaltensgenetik: eine kritische Übersicht', *Biologisches Zentralblatt*, 97, pp. 641-65.

Schilcher, F. von (1982), 'Evolution and the Nature and Nurture of Behaviour', in D. Schmidt and G. Tembrock, *Evolution and Determination of Animal and Human Behaviour*, 1982.

Schmitt, A. (1955), 'Der nominale Charakter des sogenannten Verbums der Eskimosprache', *Zeitschrift für Vergleichende Sprachforschung*, vol. 73-4, pp. 27-45.

Schrödinger, E. (1944), *What is Life*?, Cambridge University Press.

Scott, J. (1981), 'Natural Selection in the Primordial Soup', *New Scientist*, 91, pp. 153-5.

Searle, J. R. (1979a), 'Intentionality and the Use of Language', in Margalit (ed.) (1979), pp. 181-197.

Searle, J. R. (1979b), 'What is an Intentional State?', *Mind*, LXXXVIII, pp. 74-92.

Searle, J. R. (1980), 'Minds, Brains and Programs', *The Behavioural and Brain Sciences*, 3, pp. 417-57.

Seattle (1854), *Chief Seattle's Testimony*, English translation by the missionary Dr Smith of Chief Seattle's 1864 tribal oration, London 1976: Pax Christi and Friends of the Earth.

Seidenberg, M. S. and Petitto, L. A. (1979), 'Signing Behaviour in Apes: A Critical Review', *Cognition*, 7, pp. 177-215.

Seyfarth, R. M., Cheney, D. L. and Marler, P. (1980), 'Vervet Monkey Alarm Calls: Semantic Communication in a Free-Ranging Primate', *Animal Behaviour*, 28, pp. 1070-94.

Sherfey, M. J. (1972), *The Nature and Evolution of Female Sexuality*, New York: Random House.

Simon, H. A. (1955), 'A Behavioural Model of Rational Choice', *Quarterly Journal of Economics*, 69, pp. 99-118.

Simon, H.A. (1962), 'The Architecture of Complexity', *Proceedings of the American Philosophical Society*, 106, pp. 467-82; reprinted in *The Sciences of the Artificial*, Cambridge, Mass.: MIT Press.

Simpson, G. G. (1949), *The Meaning of Evolution*, New Haven, Conn.: Yale University Press.

Singer, P. (1981), *The Expanding Circle*, Clarendon Press: Oxford.

Sipes, R. G. (1973), 'War, Sports and Aggression: An Empirical Test of Two Rival Theories', *American Anthropologist*, 75, pp. 64-86.

Skinner, B. F. (1971), *Beyond Freedom and Dignity*, New York: Knopf.

Slobin, D. (1973), 'Cognitive Prerequisites for the Development of Grammar', in C. Ferguson and D. Slobin (eds), *Studies of Child Language Development*, New York: Holt, Rinehart, Winston.

Smith, F. and Miller, G. A. (eds) (1966), *The Genesis of Language – A Psycholinguistic Approach*, Cambridge, Mass., MIT Press.

Sober, E. (1975), *Simplicity*, Oxford: Clarendon Press.

Sober, E. (1980), 'Holism, Individualism and the Units of Selection', *Proceedings of Science Association*, vol. 2.

Sommers, F. (1978), 'The Grammar of Thought', *Journal of Social and Biological Structures*, 1, pp. 39-51.

Sperry, R.W., Zaidel, E. and Zaidel, D. (1979), 'Self-recognition and social awareness in the deconnected minor hemisphere', *Neuropsychologia*, 17, pp. 153-66.

Spiegelman, S. *et al.* (1968), 'The mechanism of RNA–replication', *Cold Spring Harbor Symposium*, vol. 23, pp. 101-24.

Spiro, M. E. (1979), *Gender and Culture: Kibbutz Women Revisited*, Durham NC: Duke University Press.

Sprigge, T. (1979), 'Metaphysics, Physicalism and Animal Rights', *Inquiry*, 22, pp. 101-43.

Stanley, S. M. (1979), *Macroevolution, Pattern and Process*, San Francisco: W. H. Freeman & Co.

Stanley-Peters, P. (Jr), and Ritchie, R. W. (1971), 'On Restricting the Base Component of Transformational Grammars', *Information and Control*, 18, pp. 483-501.

Stanley-Peters, P. (Jr), and Ritchie, R. W. (1973), 'On the Generative Power of Transformational Grammars', *Information Sciences*, 6, pp. 49-83.

Stebbins, G. L. and Ayala, F. J. (1981), 'Is a New Evolutionary Synthesis Necessary?', *Science*, 213, pp. 967-71, 1981.

Stegmüller, W. (1969, 1975), *Haupströmungen der Gegenwartsphilosophie*, Stuttgart: Kröner, 2 vols.

Stegmüller, W. (1977), 'The So-Called Circle of Understanding' in *Collected Papers on Epistemology, Philosophy of Science and History of Philosophy*, vol. II, pp. 3-25, Dordrecht: Reidel.

Steklis, H. D. and Raleigh, M. J. (eds) (1979), *Neurobiology of Social Communication in Primates*, New York: Academic Press.

Stent, G. S. (1978), A review of *The Selfish Gene* by Richard Dawkins, Hastings-on-Hudson, New York, Hastings Institute, 1978.

Stent, G. S. (1980). 'The Genetic Approach to Developmental Neurobiology', *TINS*, February, pp. 49-51.

Stephenson, P. H. (1979), 'A Note on the Dialectical Evolution of Human Communications Systems', *Journal of Human Evolution*, 8, pp. 581-3.

Stern, C. (1973), *Principles of Human Genetics*, San Francisco: W. H. Freeman & Co.

Suarez, S.D. and Gallup, G.G. (1981), 'Self-Recognition in Chimpanzees and Orangutans, but not in Gorillas', *Journal of Human Evolution*, 10, pp. 175-88.

Symons, D. (1979), *The Evolution of Human Sexuality*, New York: Oxford University Press.

Teilhard de Chardin, P. (1959), *The Phenomenon of Man*, New York: Harper & Row.

Tennant, N. (1978), *Natural Logic*, Edinburgh University Press.

Tennant, N. (1983), 'Evolutionary v. Evolved Ethics', *Philosophy*.

Terrace, H. S., Petitto, L. A., Sanders, R. J., and Bever, T. G. (1979), 'Can an Ape Create a Sentence?', *Science*, vol. 206, no. 4421, pp. 891-902, 23 November.

Thoday, J. M. (1953), 'Components of Fitness', *Symp. Soc. Exp. Biol.*, 7, pp. 96-112.

Thomas, R.K. (1982), 'The Assessment of Primate Intelligence', *Journal of Human Evolution*, 11, pp. 247-55.

Tiler, L. and Marslen-Wolson, W. (1981), 'Quick on the Uptake', *New Scientist*, pp. 608-9, 5 March.

Tremaine, R.V. (1975) 'Piagetian Equilibrium Processes in Syntax Learning', in Dato (ed.) (1975).

Trevarthen, C. (1979), 'Instincts for Human Understanding and for Cultural Cooperation: Their Development in Infancy', in M. Von Cranach *et al.* (eds) (1979).

Trivers, R. L. (1971), 'The Evolution of Reciprocal Altruism', *Quarterly Review of Biology*, 46, pp. 35-57, 1971.

Trivers, R. L. (1972), 'Parental Investment and Sexual Selection', in R. Campbell (ed.) (1974).

Trivers, R. L. (1974), 'Parent-Offspring Conflict', *American Zoologist*, 14, pp. 249-64.

Trivers, R. L. and Willard, D. E. (1973), 'Natural Selection of Parental Ability to Vary the Sex Ratio of Offspring', *Science*, 179, pp. 90-2.

Trivers, R.L. and Hare, H. (1976), 'Haplodiploidy and the Evolution of the Social Insects', *Science*, 191, pp. 249-63.

Turnbull, C. (1972), *The Mountain People*, New York: Simon & Schuster.

Tverskey, A. and Kahneman, D. (1974), 'Judgement Under Uncertainty: Heuristics and Biases', *Science*, 185, pp. 1124-31.

Uexküll, J. (1928), *Theoretische Biologie*, Berlin.

Van Straaten, Z. (ed.) (1980), *Philosophical Subjects*, London: Oxford University Press.

Varney, N. R. and Vilensky, J. A. (1980), 'Neuropsychological Implications for Pre-Adaptation and Language Evolution', *Journal of Human Evolution*, 9, pp. 223-6.

Vauclair, J. (1982), 'Sensorimotor Intelligence in Human and Non-Human Primates', *Journal of Human Evolution*, 11, pp. 257-64.

Vollmer, G. (1975), *Evolutionäre Erkenntnistheorie*, Stuttgart: Hirzel.

Von Cranach, M., Foppa, K., Lepennies, W. and Ploog, D. (eds) (1979), *Human Ethology: Claims and Limits of a New Discipline*, Cambridge University Press.

Von Glaserfeld, E. (1976), 'The Development of Language as Purposive Behaviour', in Harnad *et al.* (eds) (1976), pp. 212-26.

Waddington, C. H. (1953), 'Genetic Assimilation of an Acquired Character', *Evolution*, 7, pp. 118-26.

Wade, M. J. (1977), 'An Experimental Study of Group Selection', *Evolution*, 31, pp. 134-153.

Wade, N. (1980), 'Does Man Alone Have Language? Apes Reply in Riddles, and a Horse Says Neigh', *Science*, 208, pp. 1349-51, 20 June.

Walker, L. C. (1981), 'The Ontogeny of the Neural Substrate for Language', *Journal of Human Evolution*, 10, pp. 429-41.

Wallace, B. (1977), 'Automatic Culling and Population Fitness', *Evolutionary Biology*, 10, pp. 265-76.

Wasow, T. (1978), 'On Constraining the Class of Transformational

Languages', *Synthese*, 39, pp. 81-104.

Webster, G. and Goodwin, B.C. (1982), 'The Origin of Species: A Structuralist Approach', *Journal of Social and Biological Structures*, 5, pp. 15-41.

Weismann, A. 'Die Kontinuität des Keimplasmas als Grundlage einer Theorie der Vererbung', in *Aufsatze über Vererbung und verwandte biologische Fragen*, Jena: Fischer.

Weizsäcker, C. F. von (ed.) (1974), 'Evolution und Entropiewachstum', in E. von Weizsäcker (ed.), *Offene Systeme I*, Stuttgart: Ernst Klett.

Weizsäcker, E. von (ed.) (1974), *Offene Systeme I*, Stuttgart: Ernst Klett.

White, A. D. (1896), *A History of the Warfare of Science with Theology in Christendom*, New York (published also by Dover, 1960).

White, L. A. (1959), *The Evolution of Culture: The Development of Civilisation To the Fall of Rome*, New York: McGraw-Hill.

Wickler, W. (1971), *Die Biologie der Zehn Gebote*, Munich: Piper.

Willerman, L. (1979), *The Psychology of Individual and Group Differences*, San Francisco: W. H. Freeman & Co.

Williams, G.C. (1966), *Adaptation and Natural Selection*, Princeton University Press.

Williamson, P. G. (1981), 'Morphological Stasis and Developmental Constraint: Real Problems for Neo-Darwinism', *Nature*, 294, pp. 214-15.

Wilson, D. S. (1980), *The Natural Selection of Populations and Communities*, Menlo Park: Benjamin/Cummings.

Wilson, E. O. (1971), *The Insect Societies*, Cambridge, Mass.: Harvard University Press.

Wilson, E. O. (1975), *Sociobiology: The New Synthesis*, Cambridge, Mass.: Harvard University Press.

Wilson, E. O. (1978), *On Human Nature*, Cambridge, Mass.: Harvard University Press.

Winch, P. (1958), *The Idea of a Social Science*, London: Routledge & Kegan Paul.

Wittgenstein, L. (1953), *Philosophical Investigations*, Oxford: Blackwell.

Woodruff, G. and Premack, D. (1979), 'Intentional Communication in the Chimpanzee: The Development of Deception', *Cognition*, 7, pp. 333-62.

Wright, S. (1932), 'The Roles of Mutation: In-Breeding, Cross-Breeding, and Selection Evolution', *International Congress of Genetics*, 1, pp. 356-366.

Wright, S. (1945), 'Tempo and Mode in Evolution: a Critical Review', *Ecology*, 26, pp. 415-19.

Wuketits, F. M. (1981), *Biologie und Kausalität*, Berlin: Paul Parey.

Wynn, T. (1981), 'The intelligence of Oldowan hominids', *Journal of Human Evolution*, 10, pp. 529-41.

Wynne-Edwards, V. C. (1962), *Animal Dispersion in Relation to Social Behaviour*, Edinburgh: Oliver & Boyd.

Zahavi, A. (1975), 'Mate Selection – a Selection for a Handicap', *Journal of Theoretical Biology*, 53, pp. 205-14.

Ziff, P. (1981), 'Art and Sociobiology', *Mind*, XC, pp. 505-20.

NAME INDEX

SUBJECT INDEX